MA

D0248132

PHYSICAL GEOGRAPHY

Physical Geography: The Key Concepts is a thought-provoking and up-to-date introduction to the central ideas and debates within the field. It provides extended definitions of terms that are fundamental to physical geography and its many branches, covering topics such as:

- biogeography
- ecology
- climatology
- meteorology
- geomorphology
- hydrology
- pedology.

Complete with informative tables, diagrams, and suggestions for further reading, this is a highly accessible guide for those studying physical geography and related courses.

Richard Huggett is a Reader in Physical Geography at the University of Manchester. His publications include *Fundamentals of Biogeography*, *Fundamentals of Geomorphology* and *The Natural History of the Earth*, all published by Routledge.

700035929832

ALSO AVAILABLE
FROM ROUTLEDGE

The Complete Guide to Climate Change
Brian Dawson and Matt Spannagle
978–0–415–47790–1

Companion Encyclopedia of Geography
Ian Douglas, Richard Huggett and Chris Perkins
978–0–415–43169–9 (2 vols)

Fundamentals of the Physical Environment (4th edn)
Peter Smithson, Ken Addison, Ken Atkinson
978–0–415–39516–8

Fifty Key Thinkers on the Environment
Joy Palmer, David Cooper and Peter Blaze Corcoran
978–0–415–14699–9

Fifty Key Thinkers on Development
David Simon
978–0–415–33790–8

PHYSICAL GEOGRAPHY

The Key Concepts

Richard Huggett

Routledge
Taylor & Francis Group

LONDON AND NEW YORK

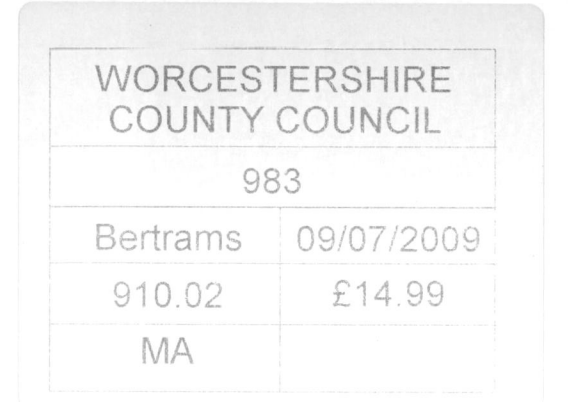

WORCESTERSHIRE
COUNTY COUNCIL

983	
Bertrams	09/07/2009
910.02	£14.99
MA	

First published 2010
by Routledge
2 Park Square, Milton Park, Abingdon, Oxon OX14 4RN

Simultaneously published in the USA and Canada
by Routledge
270 Madison Ave, New York, NY 10016

Routledge is an imprint of the Taylor & Francis Group, an informa business

© 2010 Richard Huggett

Typeset in Bembo by
Book Now Ltd, London
Printed and bound in Great Britain by
TJ International Ltd, Padstow, Cornwall

All rights reserved. No part of this book may be reprinted or
reproduced or utilised in any form or by any electronic,
mechanical, or other means, now known or hereafter invented,
including photocopying and recording, or in any information
storage or retrieval system, without permission in writing from the publishers.

British Library Cataloguing in Publication Data
A catalogue record for this book is available from the British Library

Library of Congress Cataloging in Publication Data
Huggett, Richard J.
Physical geography : the key concepts / Richard Huggett.
p. cm.—(Routledge key guides)
Includes bibliographical references and index.
1. Physical geography. I. Title.

GB54.5.H84 2009
910′.02–dc22   2008054417

ISBN10: 0–415–45207–4 (hbk)
ISBN10: 0–415–45208–2 (pbk)
ISBN10: 0–203–87567–2 (ebk)

ISBN13: 978–0–415–45207–6 (hbk)
ISBN13: 978–0–415–45208–3 (pbk)
ISBN13: 978–0–203–87567–4 (ebk)

For my family

CONTENTS

Preface   ix
Acknowledgements   xi
List of Key Concepts   xii

KEY CONCEPTS   1

Bibliography   180
Index   206

PREFACE

Physical geography is a diverse discipline. I doubt that few of its practitioners today would care to describe themselves as physical geographers except when wishing to distinguish themselves from human geographers. Most specialize in a branch of physical geography – biogeography, climatology and meteorology, ecology, geomorphology, hydrology, or pedology. The chief aim of this book is to provide extended definitions of concepts and terms that are central to discourse within physical geography and its many branches, and that are helpful for undergraduate students and lay readers. All entries will include a clear and full definition of the concept. Some of the entries for more controversial topics, such as 'uniformitarianism', will also include a short critical appraisal of the concept itself.

In selecting terms, I have used three criteria: first, they are germane to physical geography as a whole (e.g. 'energy', 'equilibrium', 'feedback'); second, they are central to a branch of physical geography (e.g. 'dispersal', 'etchplanation', 'natural selection'); third, they are important concepts from other disciplines that play a starring role in some aspect of physical geography (e.g. 'plate tectonics'). It proved exceedingly difficult to select a hundred or so concepts that I felt were 'key' to the discipline, and I am acutely aware of concepts that, for want of wordage, are excluded. Indeed, I suspect that my physical geographical peers will not think all my chosen concepts are key, and that they can offer other concepts that they would regard as key. Perhaps that says something about the rich diversity of physical geography and its practitioners. Nonetheless, I trust that all readers will find something of interest in the discussions on offer, and will have as much pleasure in reading the material as I had in researching and writing it.

I should like to thank those people who have made the completion of this book possible: Nick Scarle for drawing the diagrams; Ian Douglas for kindly commenting on my original list of key concepts; Andrea Harthill, formerly of Routledge, for asking me to write the book; and David Avital and Katherine Ong of Routledge, for help during the later stages of writing. As always, special thanks go to my wife and to my two youngest children for letting me use my PC occasionally.

Richard Huggett
Poynton
November 2008

ACKNOWLEDGEMENTS

The author would like to thank Taylor and Francis Books UK for granting permission to reproduce material in this work: Figure 5 from R. J. Huggett (2007) Drivers of global change, in I. Douglas, R. Huggett, and C. Perkins (eds) *Companion Encyclopedia of Geography: From Local to Global*, pp. 75–91, Abingdon: Routledge, Figure 6.1 (p. 83); Figures 15, 18, 38, and 39 from R. J. Huggett (2007) *Fundamentals of Geomorphology*, 2nd edn, London: Routledge, Figures 1.7 (p. 16), 1.8 (p. 17), 1.9 (p. 18), and 2.2 (p. 37); Figures 16 and 35 from R. J. Huggett (2007) Climate, in I. Douglas, R. Huggett, and C. Perkins (eds) *Companion Encyclopedia of Geography: From Local to Global*, pp. 109–28, Abingdon: Routledge, Figures 8.1 (p. 111) and 8.4 (p. 121); Figures 32 and 36 from R. J. Huggett (2004) *Fundamentals of Biogeography*, 2nd edn, London: Routledge, Figures 2.3 (p. 19) and 10.7 (p. 171).

LIST OF KEY CONCEPTS

active and passive margins
actualism/non-actualism
adaptation
adaptive radiation
advection
aridity
astronomical (orbital) forcing

bioaccumulation and biomagnification
bioclimate
biodiversity and biodiversity loss
biogeochemical cycles
bombardment

carrying capacity
catastrophism
catena
chronosequence
climate change
climax community
community change
complexity
continental drift
convection
cyclicity/periodicity

desertification
directionalism
dispersal
disturbance
drainage basin

ecological niche
ecoregion
ecosystem
ecotone
endogenic (internal) forces
energy/energy flow
environment
environmental change
equifinality
equilibrium
ergodicity (space–time or location–time substitution)
etchplanation
eustasy
evolution
evolutionary geomorphology
evolutionary pedology
exogenic (external) forces
extinction

feedback
functional–factorial approach

Gaia hypothesis
general circulation of the atmosphere
general circulation of the oceans
geochronology
geodiversity
geographical cycle
geological cycle
global warming
gradualism

habitat
habitat loss and habitat fragmentation
homeostasis/homeorhesis
hydrological cycle

invasive species
island biogeography, theory of
isostasy

land degradation

landscape ecology
limiting factors and tolerance range
local climate (topoclimate)

magnitude and frequency
mass balance
microclimate

natural selection
no-analogue communities

pedogenesis
plate tectonics
plume tectonics
populations/metapopulations

refugia
region
resilience

scale
sea-level change
soil
soil–landscapes
solar forcing
speciation
succession
sustainability
systems

taxonomy
tectonics and neotectonics
teleconnections
thresholds
time
topography
transport processes

uniformitarianism

vicariance

zonality

PHYSICAL GEOGRAPHY

The Key Concepts

Malvern Library

A

A(to form sub-
du   r where they
slic   ire common
aro   e alternative
nan   orthwest of
Nor   are active
marg   fornia is an
activ   irthquakes,
volca   characteristic c   ..... where two continental
plates   , ... 1). They are not the sites of plate boundaries, and
although continental crust abuts oceanic crust, they are part of the
same tectonic plate and subduction does not take place. The east coast
of the Americas and the west coast of Africa and Europe are examples
of passive (or Atlantic-type) continental margins. Tectonic activity at
passive margins is negligible as no plate collision or subduction occurs.

The distinction between active and passive margins has proved an
influential concept in understanding many aspects of continental
geomorphology. Active margins characteristically involve mountains
(or island arcs), with short rivers and little or no continental shelf that
plunges steeply into an offshore subduction trench. Passive margins
have generally low relief (although mountains do occur), long rivers
(such as the Amazon and Mississippi), and wide continental shelves
with thick piles of sediment. Figure 2 shows the basic geomorphic
features of passive margins with mountains. The formation of the
features is uncertain, but the starting point is probably an old plain
(palaeoplain) of a continental interior that breaks along a rift valley
(Ollier and Pain 1997). The palaeoplain at the new continental edge,
created by the rifting, downwarps. Sea-floor spreading then favours
the growth of a new ocean in which post-rift sediments accumulate as
a wedge on the submerged palaeoplain to form a seawards-sloping
basal unconformity. This is the breakup unconformity owing to its
association with the fragmenting of a supercontinent (Ollier 2004).
Inland, the palaeoplain survives as plateaux. Some plateaux may be
depositional but most are erosion surfaces formed of uplifted palaeo-
plains. In areas where the sedimentary strata form folds, the uplands
are bevelled cuestas and accordant, level strike ridges. The plateaux
may extend over large areas or they may have suffered dissection and
survive as fragments on the hardest rocks. They often retain the
ancient drainage lines. Marginal swells are widespread asymmetrical
bulges along continental edges that fall directly into the sea with

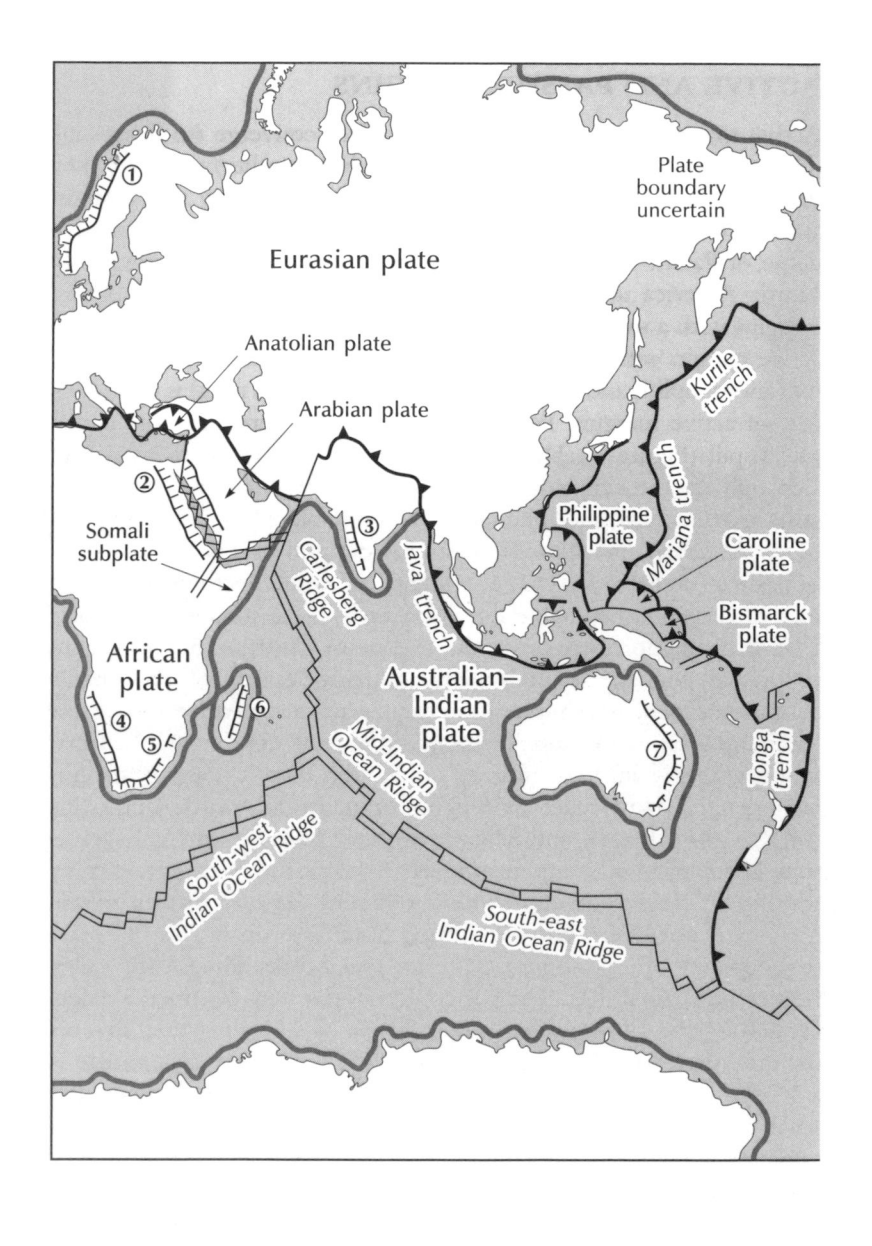

Plate boundary uncertain

Eurasian plate

Anatolian plate

Arabian plate

Kurile trench

Somali subplate

Carlesberg Ridge

Java trench

Philippine plate

Mariana trench

Caroline plate

Bismarck plate

African plate

Australian–Indian plate

Mid-Indian Ocean Ridge

Tonga trench

South-west Indian Ocean Ridge

South-east Indian Ocean Ridge

① ② ③ ④ ⑤ ⑥ ⑦

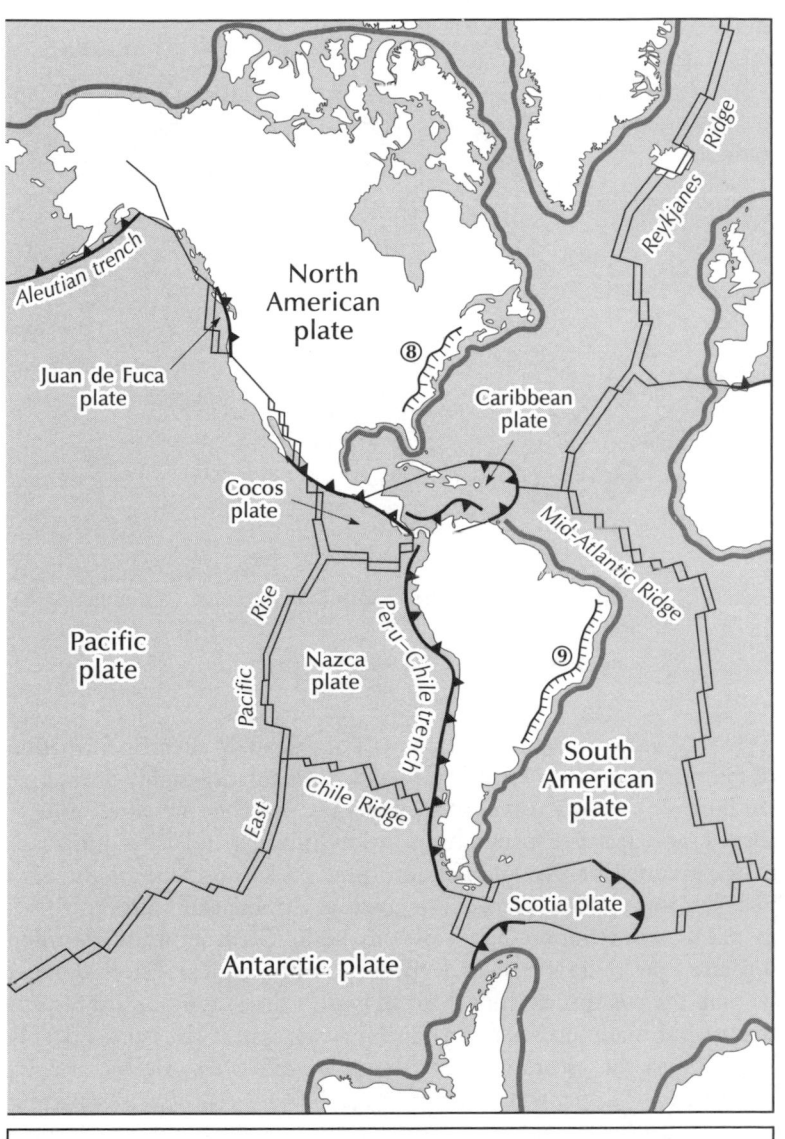

Figure 1 Tectonic plates, active and passive margins, and great escarpments.

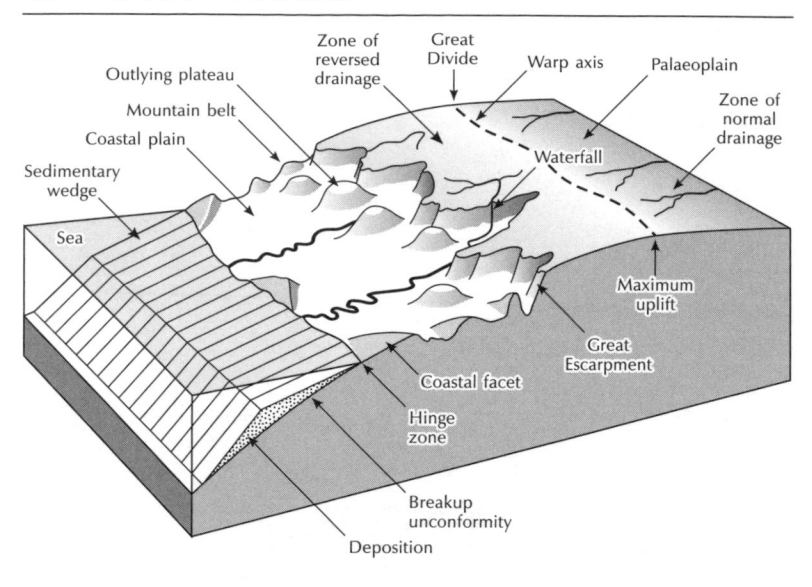

Figure 2 The chief morphotectonic features of a passive continent margin with mountains.

Source: Adapted from Ollier and Pain (1997)

steeper (2°) slopes towards the coast. They develop after the formation of plateaux and major valleys. Great escarpments are highly distinctive landforms of many passive margins (Figure 1). They are extraordinary topographic features formed in a variety of rocks (folded sedimentary rocks, granites, basalts, and metamorphic rocks) and separate the high plateaux from coastal plains. The great escarpment in southern Africa in places stands more than 1,000 m high. Great escarpments often separate soft relief on inland plateaux from highly dissected relief beyond the escarpment foot. Not all passive margins bear great escarpments, but many do, including in Norway, where the valleys deeply incised into the escarpment, although modified by glaciers, are still recognizable (Lidmar-Bergström *et al.* 2000). Some passive margins that lack great escarpments do possess low marginal upwarps flanked by a significant break of slope. The Fall Line on the eastern seaboard of North America marks an increase in stream gradient and in places forms a distinct escarpment. Below great escarpments, rugged mountainous areas form through the deep dissection of old plateaux surfaces. Many of the world's large waterfalls lie where a river crosses a great escarpment, as in the Wollomombi Falls, Australia. Lowland or coastal plains lie seawards of great escarpments. They are largely the products of erosion. Offshore from the coastal plain is a wedge of sediments, at the base of which is an unconformity, sloping seawards.

Further reading: Ollier 2004; Summerfield 1991

ACTUALISM/NON-ACTUALISM

Actualism is the supposition that no biological and geological processes other than those seen in operation today have operated in the past when circumstances were different. This belief, sometimes called the uniformity of process, was an integral part of Charles Lyell's geological credo. Lyell was convinced that, with the sole exception of the Creation, ordinary processes of Nature seen in action at present could explain all past events. He opined that when geological phenomena defy explanation in terms of present processes, then ignorance of the terrestrial system is to blame, and the invocation of processes no longer in operation is unnecessary. Non-actualism is the polar alternative idea that some past processes do not operate today.

In nineteenth-century geological circles, Lyell and other uniformitarians held staunchly to the principle of uniformity of process. However, advocates of **catastrophism** were ambivalent about it, generally agreeing that present processes should be used to explain past events whenever possible, but being fully prepared to invoke, if necessary, processes that no longer operated. In fact, the dividing line between actualists and non-actualists was not always hard and fast. Georges Cuvier, for instance, was of the firm opinion that the powers now acting at the surface of the Earth are insufficient to produce the past revolutions and catastrophes recorded in the crust. On the other hand, the English school of catastrophists – Daniel Conybeare, Adam Sedgwick, and William Buckland – all believed that the same physical causes (processes) as those in operation at present could also explain the phenomena of the past, and that the same physical laws describe the slow and gentle changes as well as the sudden and violent ones.

Non-actualistic beliefs did not vanish with the rise to supremacy of **uniformitarianism** during the nineteenth century; they just went out of fashion and lurked in the background. Today, non-actualism is making a comeback, both in Earth science and in palaeoecology. Some geologists and geomorphologists are coming round to the view that the circumstances under which processes acted in the past were different. It is probably true to say that most geologists and geomorphologists today would not hesitate in applying physical and chemical laws to past situations. They would accept that the principles of sedimentation must have remained unchanged throughout Earth history – the physical and chemical weathering of pre-existing rocks; the mechanical transport of these fragments by fluids or gases, or their chemical transport in solution; and the final deposition of the sediments under gravitational settling or chemical precipitation has always obeyed the same laws. However, they would concede that, owing to irreversible changes in the state of the atmosphere, oceans, and crust, some of the parameters in those laws have altered, and that because of this present day geological and geomorphic phenomena are not the quite the same as their earlier counterparts (Table 1). Modern sediments, for example, are very different from early Precambrian sediments (Cocks and Parker 1981, 59). Indeed, a primary thrust of modern research into Precambrian strata tries to identify how the early Earth differed from the current order of Nature. It seems safe to conclude with Harold G. Reading that:

> The present is not a master key to all past environments although it may open the door to a few. The majority of past environments differ in some respect from modern environments. We must therefore be prepared, and have the courage, to develop non-actualistic models unlike any that exist today.
>
> (Reading 1978, 479)

Table 1 A rough-and-ready guide to non-actualistic divisions of Earth history

Characteristics	Time (billions of years ago)				
	4.6–4.0	*4.0–2.0*	*2.0–0.4*	*0.4–0.1*	*0.1–0*
Water	No	Yes	Yes	Yes	Yes
Life in water	No	Yes	Yes	Yes	Yes
Oxygen in atmosphere	No	No	Yes	Yes	Yes
Life on land	No	No	No	Yes	Yes
Grasses	No	No	No	No	Yes

Note: If these divisions should be valid, then processes now seen operating at the Earth's surface cannot be a key to all past exogenic phenomena, but only to those formed during the last 100 million years. However, in the same way that modern endogenic processes may be used to aid our interpreting ancient crustal phenomena, modern exogenic processes can be used as a guide to our explaining the surface features of the Earth in all 'pre-actualistic' stages, providing it is understood that the context in which the processes operate has altered.

Source: Adapted from Huggett (1997b, 148)

Some studies of past communities by palaeoecologists also have a non-actualistic element. The discovery of **no-analogue communities** suggests climatic conditions that do not exist today. From about 18,000 to 12,000 years ago in north-central USA, a boreal grassland community rich in spruce and sedges thrived (Rhodes 1984). It occupied a broad swath of land south of the ice sheet and has no modern counterpart, though it bore some resemblance to the vegetation found in the southern part of the Ungava Peninsula, in northern Quebec, Canada, today. Its presence is due to the climate in that region being characterized by heightened seasonality and springtime peaks in solar radiation, which conditions occur nowhere at present.

Further reading: Huggett 1997b

ADAPTATION

The concept of adaptation is central to biology, and especially to evolutionary biology. Most hereditary features of organisms confer an advantage to life in a particular **environment**. Such features are adaptive – they are an adaptation that has resulted from **natural selection**. Woodpeckers possess a suite of adaptive characters that enable them to occupy their niche – chisel bill, strong head bones and head muscles, and extensile tongue with a barbed tip, feet with sharp-pointed toes pointing forwards and backwards to aid clinging to tree trunks, and a stout tail to prop up the body while clinging. Most organisms have general and special adaptations. General adaptations fit the organism for life in a broad environmental zone – a bird wing is an example. Special adaptations allow for a specialized way of life, as with the chisel bill and clinging foot of woodpeckers. It is possible that organisms possess characters that are non-adaptive or neutral, but this point is debatable.

The science of ecomorphology studies the relationships between the ecological roles of individuals and their morphological (form) adaptations, and the science of ecophysiology (or physiological ecology) delves into relationships between the ecological roles of individuals and their physiology. The life-forms of organisms commonly reflect these structural and physiological adaptations. An organism's life-form is its shape or appearance, its structure, its habits, and its kind of life history. It includes overall form (such as herb, shrub, or tree in the case of plants), and the form of individual features (such as leaves). Importantly, the dominant types of plant in each ecological zone tend to have a life-form finely tuned for survival under that climate. A widely used classification of plant life-forms, based on the position of the shoot-apices (the tips of branches) where new buds appear, was designed by Christen Raunkiaer in 1903 and distinguishes five main groups: therophytes, cryptophytes, hemicryptophytes, chamaephytes, and phanerophytes (see Raunkiaer 1934). Animal life-forms, unlike those of plants, tend to match taxonomic categories rather than ecological zones. For example, most mammals are adapted to, and their life-forms classified in accordance with, basic **habitat** types. They may be adapted for life in water (aquatic or swimming mammals such as whales and otters), underground (fossorial or burrowing mammals such as gophers), on the ground (cursorial or running and saltatorial or leaping mammals, such as horses and jerboas, respectively), in trees (arboreal or climbing mammals such as lemurs), and in the air (aerial or flying mammals such as bats).

Various organisms display remarkable adaptations to relatively extreme environments – dry, wet, hot, freezing, acidic, alkaline, and so on. Some animals and plants, for instance, have several well-known adaptations enabling them to survive in dry climates. Other organisms have adaptations enabling them to survive in the very harshest of environments. These extremophiles include hyperthermophiles (adapted to very hot environments), psychrophiles or cryophiles (adapted to very cold environments), and halophiles (adapted to salty environments) (Gerday and Glansdorff 2007). Adaptations to middle-of-the-road environments can be subtle. An example is adaptation to gradual geographical changes in climate across continents. Such adaptation often expresses itself in the phenotype (the observed characteristics of a species, resulting from the expression of the genotype interacting with the environment) as a measurable change in size, colour, or some other trait. The gradation of form along a climatic gradient is a cline (Huxley 1942). Biogeographical rules reflect clinal variation, as in Bergmann's Rule, which captures the general tendency of larger forms within a species to live in colder parts of the species' range. Other biogeographical rules relate to clines in pigmentation and the size of body extremities (such as ears) (see Huggett 2004, 16).

The concept of adaptation seems easy and commonsensical, but it is one of the most bothersome and mystifying concepts in natural history. This is especially so when considering the origins of adaptations. Feathers are now an adaptation for flying, but they evolved before birds were adept fliers, so what use were they then? The answer to riddles such as this may lie in changes of function – early 'flightless' wings might have functioned as stabilizers for fast-running birds or perhaps as heat regulators. Exaptation is a process conjectured to lie behind such changes of function, whereby characters acquired from ancestors are co-opted for a new use. An example is the blue-tailed gliding lizard (*Holaspis guentheri*) from tropical Africa that has a flattened head, which allows it to hunt and hide in narrow crevices beneath bark, and also allows it to glide from tree to tree. The head flattening was originally an adaptation to crevice use that was later co-opted for gliding (an exaptation) (Arnold 1994).

Further reading: Rose and Lauder 1996; Willmer *et al.* 2004; Gerday and Glansdorff 2007

ADAPTIVE RADIATION

Adaptive radiation is the diversification of species to fill a wide variety of **ecological niches**. It is one the most important processes bridging ecology and **evolution**. It occurs when a single ancestor species diverges, through repeated **speciation**, to create many kinds of descendant species that become or remain sympatric (live in the same area). These species tend to diverge to avoid competing with each other for resources (interspecific competition). Even when radiation generates allopatric species (species that live in different areas), some divergence still occurs as the allopatric species adapt to different environments. The 'tree of life' results from a grand adaptive radiation over some 4 billion years, with the main branches (kingdoms, phyla, and so on) and their sub-branches (families and genera) all undergoing individual adaptive radiations (Figure 3). The exception to this is the prokaryotes (bacteria and Archaea), where the transfer of genetic material between unrelated organisms occurs.

Examples of adaptive radiation are legion. Darwin's finches (Geospizinae) on the Galápagos Islands are a famous example. A single ancestor, possibly similar to the modern blue-black grassquit (*Volatinia jacarina*), colonized the archipelago from South America around 100,000 years ago. Allopatric **speciation** resulting from repeated episodes of colonization and divergence within the island group created 5 genera and 13 species. The beaks of the different species match their diet – seed-eaters, insect-eaters, and a bud-eater. The Hawaiian Islands, too, have nurtured several adaptive radiations. The radiation of the Hawaiian honeycreepers (Drepanidinae), originally thought to have started from a single ancestral seed-eating finch from Asia to give 23 species in 11 genera, is now known to have produced many more species in the recent past, with 29–33 recorded in historical times and 14 as subfossil remains. The radiation produced

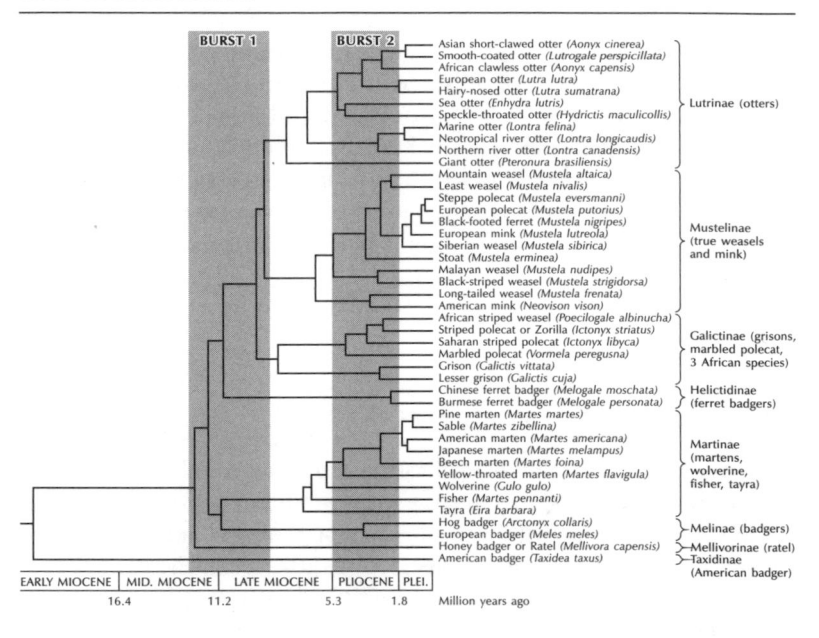

Figure 3 Adaptive radiation in the mustelids. The Mustelidae, the most species-rich family within the mammalian order Carnivora, provide a fine example of adaptive radiation. Mustelids contain 59 species classified into 22 genera and show extensive ecomorphological diversity. Different lineages have evolved in two chief bursts of diversification to fill an array of adaptive zones, from burrowing badgers to semi-aquatic otters. Mustelids are widely distributed, with multiple genera found on different continents, although they do not inhabit Madagascar, Australia, or oceanic islands.

Source: Adapted from Koepfli *et al.* (2008)

seed-eaters, insect-eaters, and nectar-eaters, all with appropriately adapted beaks. The Hawaiian silversword alliance, described as the most remarkable example of adaptive radiation in plants, displays an extreme and rapid divergence of form and physiology. The common ancestor of the silversword alliance, which split from Californian tarweeds about 13–15 million years ago, arrived in Hawaii some 4–6 million years ago. It has produced a wide range of plants that spans almost the full variety of environmental conditions found on Hawaii, with an altitudinal range from 75 to 3,750 m. The forms include acaulescent (stemless) or short-stemmed, monocarpic (flowering and bearing fruit only once before dying) or polycarpic (producing flowers and fruit several times in one season) rosette plants; long-stemmed, monocarpic or polycarpic rosette plants; trees, shrubs, and sub-shrubs; mat plants, cushion plants; and lianas.

Lemurs in Madagascar are the product of an adaptive radiation in primates that began with the arrival of a common ancestor some 50 million years ago. At least 45 species lived in the recent past, around 2,000 years ago when humans first arrived on the island; some 33 survive today in 14 genera. The true lemurs comprise five arboreal (tree-living), vegetarian species that eat fruits, flowers, and leaves. Sportive lemurs are nocturnal and move mainly by jumps. Mouse lemurs (*Microcebus*) are small (up to 60 g), run like rodents, and eat insects as well as fruits. The indri and sifakas (*Propithecus*) are large animals (up to 1 m long). The aye-aye specializes in prising insect larvae from tree bark and fills the niche of woodpeckers. At least 15 species of subfossil lemur species in 8 or more genera reveal the 'big' end of the radiation. *Archaeolemur* lived on the ground and was about the size of a female baboon. The 77-kg *Megaladapis* was arboreal with a niche similar to that of a koala. At 60 kg, *Palaeopropithecus* was a sloth-like tree-dweller.

Not all **adaptations** are radiative and not all are adaptive. Non-radiative 'radiation' occurs when vacant niche space permits a sort of ecological release involving diversification but not **speciation** within a lineage. An example is 'o'hia lehua (*Metrosideros polymorpha*). This Hawaiian tree species is very diverse and has a wide range of forms. It occupies bare lowlands to high bogs, occurs as a small shrub on young lava flows and as a good-sized tree in a canopy of mature forest. Nevertheless, botanists ascribe it to a single species despite such a rich variety of forms. Non-adaptive radiation occurs where radiation is associated with no clear niche differentiation. It may occur when radiations have occurred allopatrically in fragmented **habitats**. For instance, on Crete, land snails of the genus *Albinaria* have diversified into a species-rich genus with little niche differentiation. All species occupy roughly the same or only a narrow range of habitats, but rarely do any two *Albinaria* species live in the same place.

Further reading: Givnish and Sytsma 1997; Schluter 2000

ADVECTION

Advection is the transport of any substance or conserved property (such as heat, water vapour, and salinity) in a fluid. An example of substance advection is the transport of solutes and fine sediments in river water. The flowing water carries dissolved substances and suspended particles downstream. Any fluid can advect any substance or conserved property. In meteorology and physical oceanography, advection refers to the more-or-less horizontal transport of a conserved property (heat, moisture, salinity) from one region to another. In the atmosphere, this is important in the formation of orographic clouds (clouds resulting from the forced rising of air by such topographic features as mountains) and the precipitation of water from clouds, as part of the water cycle. The advection of warm, moist air over a cold sea or land surface may produce advection fog. In Britain, advection fog forms under two main conditions: first, moist south-west winds flowing over a progressively colder sea around the west and south coasts in spring produce sea fog and coast fog; second, east winds flowing across the cold North Sea in summer produce advection fog along the east coast. In North America, advection fogs are common between May and August off Newfoundland as warm, moist air form the south cools quickly over the cold waters of the Labrador Current.

ARIDITY

Aridity relates to the dryness of an **environment**, which has an enormous influence on animals and plants. It depends upon the interplay of temperature, evaporation, the annual amount and seasonal distribution or precipitation, and **soil** factors. Regions of high aridity encourage the formation of deserts because the meagre rainfall cannot support trees or woody plants.

Deserts are regions with very low annual rainfall (less than 300 mm), scanty vegetation, extensive areas of bare and rocky mountains and plateaux, and alluvial plains. They cover about a third of the Earth's land surface. Many deserts are hot or tropical, but some polar regions, including Antarctica, are deserts because they are dry. Aridity forms the basis of classifications of deserts. Most classifications use some combination of the number of rainy days, the total annual rainfall, temperature, humidity, and other factors. In 1953, Peveril Meigs divided desert regions on Earth into three categories according to the amount of precipitation they receive. Extremely arid lands – deserts – have at least 12 consecutive months without rainfall; arid lands – also deserts – have less than 250 mm of annual rainfall; and semiarid lands have a mean annual precipitation of between 250 and 500 mm – these are mostly prairies or steppes. Modern work tends to use a fourfold classification of drylands, which cover some 41 per cent of the land surface, based on an aridity index:

$$AI = PE / P$$

where *PE* is the potential evapotranspiration and *P* is the average annual precipitation (Table 2). Many of these dryland areas face severe **land degradation**.

Table 2 Degrees of aridity defined by an aridity index[a]

Aridity type	Aridity index	Land area in category (%)
Hyperarid	<0.05	7.5
Arid	0.50–0.20	12.1
Semi-arid	0.20–0.50	17.7
Dry subhumid	0.50–0.65	9.9

[a]United Nations Environment Programme (Middleton and Thomas 1997).

Further reading: Laity (2008)

ASTRONOMICAL (ORBITAL) FORCING

The planets and their satellites jostle with one another as they orbit the Sun. This jostling leads to medium-term orbital variations occurring with periods in the range 10,000 to 500,000 years that influence Earth's climate, not by changing the total amount of solar **energy** received by the Earth during the course of a year, but by altering the seasonal and latitudinal distribution of solar energy.

Orbital variations in the 10,000–500,000-year frequency band appear to have driven climatic change during the Pleistocene and Holocene epochs. Orbital forcing has led to climatic change in middle and high latitudes, where ice sheets have waxed and waned, and to climatic change in low latitudes, where water budgets and heat budgets have marched in step with high-latitude climatic cycles. Quaternary loess deposits, **sea-level changes**, and oxygen-isotope ratios of marine cores record the 100,000-year cycle of eccentricity. The precessional cycle (with 23,000-and 19,000-year components) and the 41,000-year tilt cycle ride on the 100,000-year cycle. They, too, generate climatic changes that register in marine and terrestrial sediments. Oxygen isotope ratios (δO^{18}) in ocean cores normally contain signatures of all the Earth's orbital cycles, though the tilt cycle, as it affects seasonality, has a stronger signature in sediments deposited at high latitudes.

James Croll and Milutin Milankovitch first proposed the theory of

climate change by astronomical forcing. The ideas of Croll and Milankovitch were popular up to about 1950, but then faded from prominence. During the late 1960s and early 1970s, researchers redis-covered Milankovitch's cycle of great seasons, which saw glacial 'winters' melting into interglacial 'springs', 'summers', and 'autumns'. Evidence for a 100,000-year great cycle of climate was unearthed independently in loess sequences exposed in a quarry in Czecho-slovakia, in sea levels, and in the oxygen-isotope ratios of marine cores. Moreover, both the terrestrial and marine records attested to long periods of glacial expansion (climatic cooling) abruptly ended by rapid deglaciations (climatic warming). Later, research demonstrated that the precessional and tilt cycles explained climatic oscillations superimposed on the 100,000-year cycle. This demonstration required a finely calibrated calendar of Pleistocene events. Largely owing to the endeavours of the members of the CLIMAP (Climate: Long-range Investigation, Mapping, And Prediction) project, a suitably detailed calendar emerged that gave an accurate chronology of late Pleistocene climate. Confirmation of the Croll–Milankovitch theory was even-tually forthcoming when scientists subjected suitable sediment cores from the Indian Ocean, which recorded climatic change over the last 450,000 years, to spectral analysis (which reveals periodicity in data sets). The results of the analysis revealed cycles of climatic change at all frequencies corresponding to orbital forcings (Hays *et al.* 1976). In addition to the 23,000-year precessional cycle, a 19,000-year preces-sional cycle component was present, the occurrence of which was later demonstrated using astronomical theory. The publication of these findings convinced most scientists that the motion of the Earth around the Sun did drive the world climate system during the late Pleistocene, that orbital variations were the 'pacemaker' of the ice ages.

The astronomical theory of **climate change** became the ruling theory of many Pleistocene and Holocene climatic and environmental fluctuations. However, variations in orbital parameters do not explain all aspects of Quaternary climatic change. Elkibbi and Rial (2001) identified five challenges to the astronomical theory of ice ages. Three relate to the '100,000-year problem'. First, 100,000-year variations of insolation forced by eccentricity changes are too small (less than 1 per cent) to drive the great ice ages. Second, 100,000-year oscillations have dominated the last 900,000 years but 41,000-year oscillations dominated the late Tertiary and early Quaternary, the switch known as the mid–Pleistocene transition. The third challenge is the '400,000-year problem', which is the absence of a 413,000-year signal in oxygen isotope ratios from marine cores over the past 1.2 million years, despite that being the largest component of eccentricity forcing. Fourth, over the last 500,000 years, the length of glacial stages ranges from about 80,000–120,000 years, which variation cannot correlate linearly with insolation changes. The fifth challenge is the presence of signals for climatic cycles that appear unrelated to insolation forcing, which indicate nonlinear responses of the climate system. In addition to these five problems is the finding that a number of palaeoclimatic records, when subjected to re-examination, have a variance attributable to orbital changes never exceeding 20 per cent (Wunsch 2004).

BIOACCUMULATION AND BIOMAGNIFICATION

Bioaccumulation occurs when an organism absorbs a toxic substance at a rate greater than it excretes it. The longer the substance is in an organism, the greater the risk to health by accumulation to harmful concentrations. Substances that accumulate in this way include methyl mercury, some lead compounds, selenium, and DDT (a chlorinated hydrocarbon), all of which are soluble in lipids (fats). Some such substances are more soluble in fats than they are in water and become concentrated in fatty tissues. For species at the bottom end of food webs, concentrations may remain low enough not to cause health problems. However, the concentrations may build up (or become magnified) as organisms eat each other and the substances move along the feeding levels. This process is biomagnification (biological magnification, bioamplification, food–chain concentration). Rachel Carson brought its inimical effects to public notice in her book *Silent Spring* (1962). This book drew attention to the alarming build-up of long-lasting pesticides, mainly DDT, in the **environment** and the damage that they were causing to wildlife and humans near the top of food chains. The insecticidal properties of DDT, first discovered by Paul Müller in 1939, led to its promotion as a complete solution to pest control. By the early 1960s, its persistence in the environment and accumulation in the food chain were becoming apparent. In 1972, after years of forceful lobbying and petitioning in the USA, DDT was banned by the Environmental Protection Agency for all but emergency use.

Some heavy metals are also stored in body tissues and are subject to bioaccumulation and biomagnification. In Louisiana, USA, animals and plants living in ditches along busy roads have accumulated cadmium and lead (Naqvi *et al.* 1993). In the red swamp crayfish (*Procambarus clarkii*), the cadmium level was 32 times that in the water, and the lead level 12 times that in the water, giving bioaccumulation factors of 5.1 and 1.7, respectively. In San Francisco Bay, evidence suggests that cadmium is biomagnified 15 times over two trophic levels within two food webs (Croteau *et al.* 2005). In the Alto Paraguay River Basin, Brazil, large quantities of mercury, used in gold mining, disperse directly into the air and into the rivers feeding the Pantanal, a wildlife reserve (Hylander *et al.* 1994). Local, commercially important catfish (*Pseudoplatystoma coruscans*) had mercury contents above the limit for human consumption, and significantly above the natural background level. Mercury content in bird feathers also indicated biomagnification. However, **soil** and sediment samples had no statistically significant accumulation of mercury. Evidently, organisms more readily absorb mercury originating from the gold-mining process than they do mercury naturally present in soil minerals.

BIOCLIMATE

Bioclimatology deals with the relationships between climate and living organisms, including humans. It focuses upon the interactions between the biosphere and the Earth's atmosphere over seasons or years (as opposed to biometeorology, which looks at daily and monthly interactions). Bioclimates depend upon climatic factors of particular relevance to life. Life's master **limiting factors** are precipitation and temperature, and these form the core of bioclimatic systems. As a rule, in tropical areas, temperatures are always high enough for plant growth and precipitation is the limiting factor; in cold environments, water is usually available for plant growth for most of the year – low temperatures are the limiting factor. This is true, too, of limiting factors on mountains where heat or water (or both) set lower altitudinal limits, and lack of heat sets upper altitudinal limits. There are several schemes for characterizing bioclimates in broad terms. A popular scheme is the 'climate diagram' devised by Heinrich Walter (e.g. Walter and Lieth 1960–67), which portrays climate as a whole, including the seasonal round of precipitation and temperature.

A recent development in bioclimate studies is the building of relatively sophisticated bioclimatic models either to predict future climatic changes on animals and plants or to infer past climates from the distribution and composition of past faunas and floras. Knowledge of the bioclimatic requirements of a species allows the prediction of its potential fate under **climate change**. Predicted changes in bioclimatic envelopes of three British species under a worst-case scenario climate model suggest a potential expansion of the great burnet (*Sanguisorba officinalis*) range, no change in the range of the yellow-wort (*Blackstonia perfoliata*), and a contraction in the range of the twin-flower (*Linnaea borealis*) by the 2050s (Berry *et al.* 2002; see also Berry *et al.* 2003). Other bioclimatic modelling studies predict the future distribution of tree species in Europe (Thuiller 2003) and identify the environmental limits for vegetation at biome and species scales in the fynbos **biodiversity** hotspot in South Africa, with a view to predicting the likely shrinkage under a warming climate (Midgeley *et al.* 2002). Cyrille Rathgeber *et al.* (2005) developed a bioclimatic model to explore the relationships between tree radial growth and bioclimatic variables closely related to the biological functioning of a tree, using Aleppo pine (*Pinus halepensis*) as a case study. The results showed that soil water availability during the growing season is the chief determinant of Aleppo pine growth, and that actual evapotranspiration (AET) is the bioclimatic variable that best expresses the observed inter-annual tree growth variations. The convincing results obtained by the bioclimatic model, as well as the limited numbers of parameters it requires, demonstrate the feasibility of using it to explore future climatic change impacts on Aleppo pine forests.

Some scientists use bioclimatic models to infer past climatic distributions from community composition, so providing a new method of palaeoclimatic reconstruction. One study explored the usually strong correlation between climate and mammal community composition (Fernández and Peláez-Campomanes 2003). By establishing relationships between worldwide modern mammal faunas and climate, a bioclimatic model was developed that was then applied to Quaternary faunas from Eurasia. More than 90 per cent of the localities were classified correctly by the model.

BIODIVERSITY AND BIODIVERSITY LOSS

Walter G. Rosen coined the term biodiversity as shorthand for biological diversity in 1985 for a National Forum on Biodiversity held in Washington in September 1986 (Wilson 1988). The ecological community swiftly adopted it and it came to public notice in 1992 when 150 nation states signed the Convention on Biological Diversity (CBD) at the United Nations Conference on the Environment and Development in Rio de Janeiro (the 'Earth Summit'). This act placed the concept of biodiversity at the head of the conservation agenda where it has stayed as the key theme for global and local conservation efforts (Lévêque and Mounolou 2003; Ladle and Malhado 2007).

Despite its popularity, the meaning of the term 'biodiversity' has remained exasperatingly difficult to pin down, with Delong (1996) listing 85 different definitions. Article 2 of the CBD perhaps gives the most widely accepted definition: 'the variability among living organisms from all sources including, *inter alia*, terrestrial, marine and other aquatic **ecosystems** and the ecological complexes of which they are part; this includes diversity within species, between species and of ecosystems'. Biodiversity thus refers to the natural variety and variability among living organisms, the space they inhabit, and their interactions with each other and the physical **environment** (Gaston and Spicer 2004). This natural biodiversity contrasts with patterns of biodiversity created by human influence. It is important to note that biodiversity is not a value-free term. Most definitions imply that biodiversity is inherently a 'good' thing, and that, in consequence, biodiversity loss through human action is a 'bad' thing and to be prevented or kept to a minimum level (Ladle and Malhado 2007).

Many commentators claim that the current 'biodiversity crisis' differs from the previous crises in the history of life (mass **extinctions**) because it is running at a uniquely rapid rate and because it is the direct result of human actions. Erosion of biodiversity occurs at various scales, from the genetic diversity of many natural and domesticated species to the diversity of whole ecosystems and landscapes. Current human-induced rates of species extinctions may be about 1,000 times greater than past background rates. Biodiversity loss is a matter of concern, not only because of the aesthetic, ethical, or cultural values attached to biodiversity, but also because it could have numerous far-reaching and often unanticipated consequences for the ecosphere as a whole. It may weaken the ability of natural and managed ecosystems to deliver ecological services such as production of food and fibre, carbon storage, nutrient cycling, and to resist climatic and other **environmental changes**. The assessments of the causes and consequences of biodiversity changes, and the institution of the foundations for the conservation and sustainable use of biodiversity, are foremost scientific challenges.

Several factors drive current biodiversity change at a global scale. There are four main groups of drivers: land-cover change, species exploitation and exchange (the accidental or deliberate introduction of plants and animals to ecosystems outside those in which they live), climatic change, and changes in environmental chemistry (Figure 4). Land-cover change works through **habitat loss** and **habitat fragmentation**. The intercontinental spread of species, often abetted by humans, and the overexploitation of species work directly on communities. Climate affects biodiversity by reorganization the distribution of bioclimatic space in which species and communities thrive.

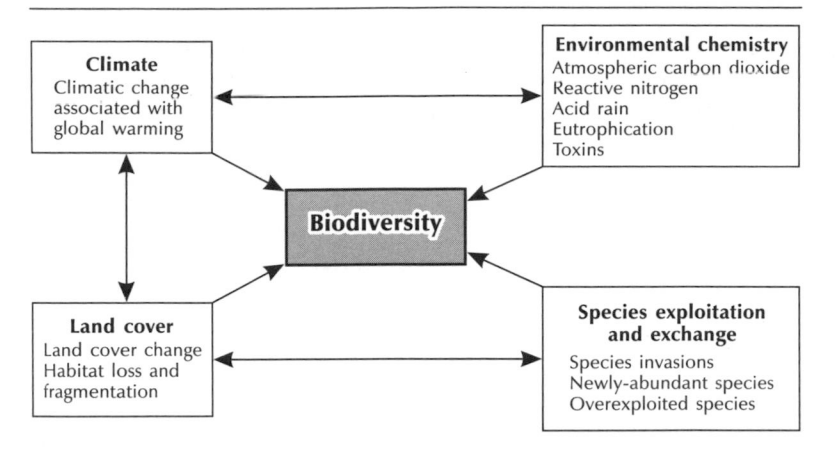

Figure 4 Drivers of biodiversity change and the main interactions between them.

Chemical changes in the environment, such as a build-up of nutrients in lakes and **soil** acidification, acts on the biosphere. These assorted factors have individual effects but they also have synergistic effects. For example, a high availability of nitrogen would augment the effects of elevated atmospheric carbon dioxide levels on ecosystems. Likewise, the effects of biotic exchanges are likely to be greater when they occur at the same time as land-use changes. Similarly, predictions suggest that the combination of **climate change** and habitat destruction can be disastrous (Travis 2003).

Further reading: Gaston and Spicer 2004; Lovejoy and Hannah 2006

BIOGEOCHEMICAL CYCLES

Biogeochemical cycles or nutrient cycles are the repeated movements of bioelements (elements essential for life) through living things, air, rocks, **soils**, and water. They all have biotic (living) and abiotic (non-living) phases, and they involve stores (pools, reservoirs, sinks) and fluxes of various chemical species in and between the atmosphere, hydrosphere, pedosphere, and lithosphere. Therefore, material exchanges between life and life-support **systems** define biogeochemical cycles. At their grandest scale, biogeochemical cycles involve the entire Earth. An exogenic cycle involves the transport and transformation of materials near the Earth's surface; a slower and less well-understood endogenic cycle involves the transport and transformation of materials in the lower crust and mantle. Bioelements such as carbon that are in a gaseous state for a leg of the cycle form gaseous biogeochemical cycles; those that do not volatilize readily and that move mainly in solution, magnesium for instance, form sedimentary biogeochemical cycles. The major cycles involve the storage and flux of hydrogen, carbon (Figure 5), nitrogen, oxygen, magnesium,

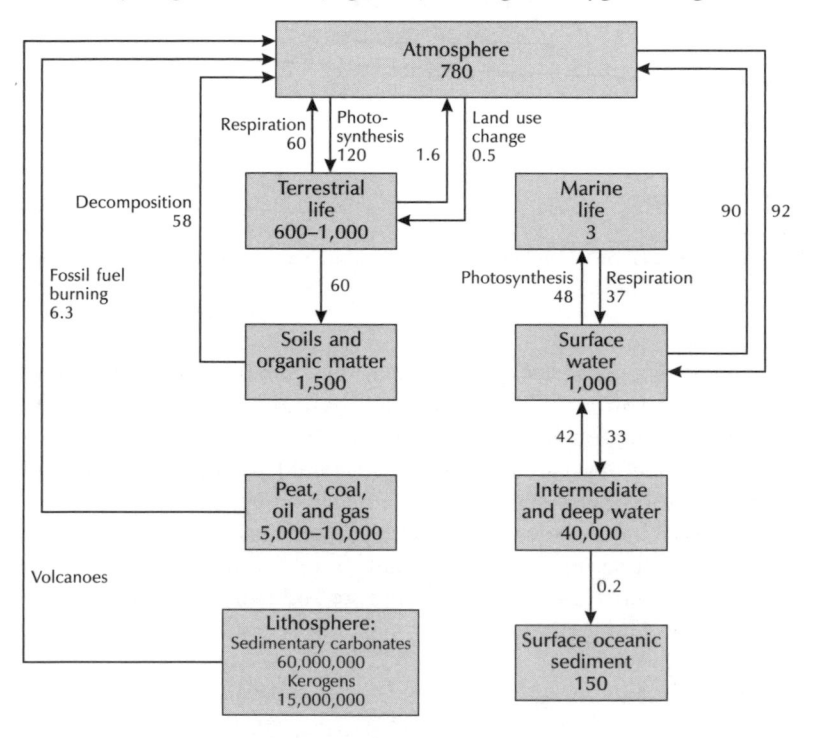

Figure 5 The carbon cycle. Stores are in gigatonnes of carbon (GtC) and flows are in GtC per year. Carbon is the basis of all life on Earth. The global carbon cycle is the return movement of carbon through living things, air, rocks, soil, and water. Photosynthesizing plants convert atmospheric carbon, in the form of carbon dioxide, to carbohydrates. Producer respiration returns some plant carbon to the atmosphere in the form of carbon dioxide. Animals assimilate and metabolize some plant carbon. A portion of animal carbon returns to the atmosphere as carbon dioxide released through consumer respiration. The balance enters the decomposer food chain and either returns to the atmosphere through decomposer respiration or accumulates as organic sediment (e.g. peat and coal). Combustion during fires and volcanic eruptions releases carbon dioxide into the atmosphere.

Source: After Huggett (2007b)

phosphorus, potassium, sulphur, and calcium. The micronutrients sodium and chlorine and many other elements suspected of being micro-bioelements (e.g. boron, molybdenum, silica, vanadium, and zinc) also have biogeochemical cycles. Humans have become a major biogeochemical force, materially altering the flows of biochemicals, which are having repercussions in the **environment**.

The carbon cycle is crucially important to life. Given the genuine threat of climatic changes associated with **global warming**, it is not surprising that much research goes into studying the effects of elevated carbon dioxide levels on **ecosystems** (e.g. Falkowski *et al.* 2000). Measurements of carbon stores and fluxes, and the understanding of carbon-cycling processes, are becoming ever more refined. Modelling studies in the early 1970s suggested that, to balance the global carbon budget, carbon must be stored in terrestrial ecosystems. Further research in the 1990s confirmed that terrestrial ecosystems play a vital role in regulating the carbon balance in the Earth system. The global carbon cycling research, so crucial to the global warming issue, has stimulated a wealth of studies on several aspects of plant biology, including leaf photosynthesis, plant respiration, root nutrient uptake, and carbon partitioning (the allocation of carbon to roots, trunks, leaves, and so on). The studies range from response to increased carbon dioxide concentrations at the molecular level, through impacts on species diversity in communities, to carbon fluxes in the ecosphere. At the global scale, global biosphere models take in experimental results to predict potential changes in terrestrial ecosystems as the globe warms up. Current research spotlights the complexity of the carbon cycle and the inadvisability of making unequivocal predictions about the impact of rising temperatures, at least for some ecosystems (Rustad 2001). Two opposing processes connected with the carbon cycle are associated with climatic warming. First, a negative **feedback** mechanism is the increase in plant growth and carbon sequestration resulting from stimulated nutrient mineralization and longer growing seasons. Second, the warming may trigger a positive feedback mechanism, stimulating biological metabolism in terrestrial vegetation and so encouraging a greater release of heat-trapping gases to the atmosphere and augmenting human-induced warming. Recognition of the second mechanism partly explains the increased temperature rises predicted by the International Panel on Climate Change (IPCC). An experiment to test this mechanism in the field, conducted in a tall-grass prairie in the Great Plains of the USA, revealed another complicating factor – acclimatization (Luo *et al.* 2001). The process here seems to be that, rather than soil respiration simply increasing with increasing temperature, the soil tends to 'acclimatize' to the higher temperatures, and does so more fully at high temperatures, so weakening the positive feedback effect. In addition, increased carbon dioxide levels increase microbial activity, which in turn promotes the formation of soil aggregates. As soil aggregates tend to protect particles of organic matter in the soil against microbial attack, an increase in soil aggregation resulting from higher carbon dioxide levels may lead to a sequestration of soil carbon. This is another negative feedback mechanism.

Further reading: Jacobson *et al.* 2000

BOMBARDMENT

Asteroids and comets strike the Earth. Nobody has witnessed a large strike in modern times, the Tunguska incident of 1908, in which a 60-m diameter meteoroid travelling at around 10 km/s exploded some 9 km above Siberia, being the largest historical event for which there is direct evidence (Gasperini *et al.* 2008). Nevertheless, space probes have shown that impact craters are common on other planets and satellites throughout the Solar System, and the **magnitude and frequency** of impact events is calculable from the size distributions of these craters. The frequency of collision with asteroids, comets, and meteoroids is inversely proportional to the size of the impacting body. Meteoritic dust continuously rains into the atmosphere; small meteoroid strikes occur almost weekly; asteroids with a diameter of about a kilometre strike about three times every million years. A recent calculation gives less than 1-in-100,000 odds on a 2–3 km diameter asteroid smiting the Earth in the next hundred years (Chapman 2004). Mountain-sized asteroids or comets, like the one that probably led to the mass extinction at the end of the Cretaceous Period, strike just once every 50 million years or thereabouts.

The immediate effect of a bolide impact is the production of a crater, sometimes called an astrobleme. It is impossible in a laboratory to replicate the processes by which large impact craters form by sudden releases of huge quantities of **energy**, and no such structure has formed during recorded human history (French 1998, 17). Researchers gain knowledge of large impact structures indirectly, by combining theoretical and experimental studies of shock waves and geological studies of larger terrestrial impact structures. All seem to agree that cratering is a complex process, which still has many uncertain details. Geological evidence of an impact crater includes shatter cones (formed by an explosion above rock strata), shock-metamorphosed forms of silica, such as coesite and stishovite, and shocked quartz crystals, all of which result from by the immense pressures created in the rocks around an impact site. Given the highly energetic nature of hypervelocity impacts, it is reasonable to speculate that they may trigger a number of geophysical processes including reversals of the Earth's magnetic field, **continental drift**, and volcanism (Napier and Clube 1979; Rampino 1989). For many years, geophysicists were sceptical about the possibility of a hefty impact triggering large-scale volcanism (e.g. Ivanov and Melosh 2003), but it seems that a giant bolide with 30-km diameter hitting a thin (75-km thick) lithosphere could produce flood-province-scale volcanism (Elkins-Tanton and Hager 2005; see also Price 2001).

A debate has raged over whether bombardment has occurred randomly or periodically through Earth history. Originally, astronomers were inclined to the idea that asteroids and comets were stray bodies, taken out of the Asteroid Belt and Oort Cloud (the vast reservoir of comets lying in the outer Solar System) by chance events. Some astronomers disputed this stray bolide hypothesis or 'stochastic **catastrophism**' (Steel *et al.* 1994, 473), favouring instead a coordinated or coherent catastrophism. Stochastic catastrophism, although demanding random strikes, does not preclude the possibility of bombardment episodes, which might also occur randomly. There are theoretical grounds, and some empirical evidence, for conjecturing that bombardment tends to occur as episodic showers ('storms' is a more apt description), roughly every 30 million years, each shower lasting a few million years (Rampino 2002). Several mechanisms might explain episodic storms of space debris. These mechanisms form the basis of three hypotheses. The Nemesis hypothesis proposes that the Sun might have a companion star on a highly eccentric orbit that perturbs the Oort Cloud at perihelion passage. The Planet X hypothesis postulates that an undiscovered tenth planet orbits in the region beyond Pluto and produces comet showers near the Earth with a very stable frequency. The 'cosmic carousel' hypothesis contends that the bobbing motion of the Solar System about the Galactic plane leads to periodic comet showers.

Other astronomers propose what might be termed a 'harmonized catastrophism' to supersede stochastic catastrophism. Two main schools advocate this new view – coherent catastrophism and coordinated catastrophism. Proponents of coherent catastrophism contend that large comets disintegrate to produce clusters of fragments, ranging in size from microns, metres, tens and hundreds of metres, to kilometres (Steel 1991, 1995; Steel *et al.* 1994). Such clusters will form a train of debris with a characteristic orbit. If the node of the orbit (the point at which it crosses the ecliptic) is near 1 AU, and if the cluster passes its node when the Earth is near, then it repeatedly crosses the Earth's orbit. The outcome is cluster–object impacts at certain times of the year, every few years, depending on the relationship between the Earth's and the cluster's orbital periods. However, an impact occurs only when precession has brought the node to 1 AU, so only on timescales of every few thousand years. One cluster – the Taurid complex – is presently active, and has been for the last 20,000 years. It has produced episodes of atmospheric detonation, which the proponents of coherent catastrophism believe may have had material consequences for the biosphere and for civilization. A caveat seems appropriate here, for not all astronomers accept this brand of cosmic catastrophism (e.g. Chapman 1996). The second brand of harmonized catastrophism is coordinated catastrophism, which sees the Earth, Sun, and Solar System as coupled nonlinear **systems** (Shaw 1994). This powerful idea leads to a new picture of Earth history that outlaws happenstance and instates chaotic dynamics as its centrepiece; a picture that shows a grand coordinated theme played out over aeons, and that portrays gradual and catastrophic change in the living and non-living worlds as different expressions of the same nonlinear processes.

Further reading: Belton *et al.* 2004; Bobrowsky and Rickman 2006; Verschuur 1998

CARRYING CAPACITY

The idea of carrying capacity assumes a balance between **populations** and resources in an **ecosystem**. Carrying capacity is the maximum species population that the resources of an **environment** or place can sustain without jeopardizing the species population, other species in the ecosystem, and all necessary ecosystem processes. If populations should exceed the carrying capacity, there will be insufficient resources to support the excess numbers. The population will then fall until a new balance prevails (at the carrying capacity). If the population is lower than the carrying capacity, then it will normally grow up to the carrying capacity. In population models, the letter K denotes carrying capacity.

Carrying capacity applies to natural ecosystems, to agroecosystems (agricultural ecosystems), and to recreational systems. In agriculture, carrying capacity determines the number of sustainable grazing animals. In recreational land–use, it determines the number of people and kinds of activity that can be accommodated without harming the environment. In all types of ecosystems, carrying capacity is as a rule constrained by such **limiting factors** as water, heat, and nutrient supply. Social and economic factors may also limit carrying capacity. Human activity often drives populations to levels above the carrying capacity and leads to environmental degradation. This may be the case when too many cattle are farmed in range management or when too many people are allowed to use a recreational area. On rangeland, the carrying capacity depends upon rainfall and **soil** fertility. In the USA, the carrying capacity of pasture and rangeland varies from about 200 cows per km^2 to 4 cows per km^2.

In theory, the planet Earth has a carrying capacity. In practice, the planetary carrying capacity can only be calculated according to an agreed set of human values – what kind of life and what kind of environment do humans want? The highest estimate would depend upon the amount of land and resources needed per person to provide necessities – food, water, clothing, and waste disposal. A lower estimate would provide a better equality of life and allow greater space for such amenities as recreation and wilderness.

CATASTROPHISM

Catastrophism, a term coined by William Whewell in 1832, is a school of thought standing in antithesis to **gradualism**. Its proponents claim that past rates of geological and biological processes differed substantially from current rates, on occasions suddenly and violently assuming magnitudes not seen today and in doing so, causing catastrophes.

In the inorganic world, it is convenient to recognize two brands of catastrophism – the old and the new. The old catastrophism was the ruling theory of Earth history before about 1830. It embodied many different ideas but a common thread running through them was the recognition of one or more global, or nearly global, revolutions in Earth history, usually associated with worldwide floods and the collapsing and crumpling of the Earth's crust. As gradualism waxed in the middle of the nineteenth century, so catastrophist views became rather disreputable. During the twentieth century, catastrophism made a comeback with some scientists again invoking catastrophic processes as agents of regional and global geological change. These catastrophes were of terrestrial origin, as in the release of large volumes of water as in the Lake Missoula Flood, or of extraterrestrial origin, as in **bombardment** by asteroids and comets. Not until 1980 did a new and acceptable brand of catastrophism emerge in the form of cosmic catastrophism. Soon after the discovery of the famous iridium layer at the Cretaceous–Tertiary boundary, it became widely accepted that bombardment by asteroids, comets, and meteorites is a plausible explanation for apparently sudden and violent events in Earth's past. Today, some researchers take the possibility that bombardment may influence **plate tectonics**, geomagnetic reversals, true polar wander, earthquakes, volcanism, climatic change, and the development of some landforms very seriously.

In the biological world, some of the old catastrophists, such as Baron Georges Cuvier (1769–1832), believed that global revolutions had exterminated nearly all organisms, so causing an overturning of biotas (animal and plant communities). In the second half of the twentieth century, some new catastrophists opined that the history of life involves abrupt and thoroughgoing changes at two levels: punctuational (catastrophic) styles of change at the level of species – sudden, but not violent, **speciation** events; and catastrophic change at the level of biotas – sudden, and probably violent, mass **extinctions**. Eldredge and Gould (1972) constructed a widely discussed punctuational model of **evolution**. They argued that large evolutionary changes condense into discontinuous speciational events (punctuations) that occur very rapidly; after a new species has evolved, it tends to remain largely unchanged. This view, they claimed, explains the pattern of species change commonly found in the fossil record. As for change of biotas, even during the reign of gradualism there were several supporters of the view that some extinction events had been sudden and violent and had resulted from truly catastrophic processes (such as outbursts of cosmic radiation) that had produced abrupt and devastating global changes. Today, several terrestrial processes, some of which act gradually, are thought capable of stressing the biosphere severely enough to induce mass extinctions, but bombardment by extraterrestrial bodies is widely accepted as a very plausible cause.

Further reading: Huggett 1997b, 2006

CATENA

In 1935, while working as a **soil** chemist in Tanganyika, Geoffrey Milne put forward the concept of the catena as a unified framework within which to study functional aspects of soils on hilly terrain. W. S. Martin, a soil chemist based in Uganda, inspired him (Brown 2006). Milne argued that a transect running from hillcrest to hollow traverses very different soil profiles that 'are not, properly speaking, individual soils at all, but are a compound soil unit of another kind in which a chain of profile-differences occurs in a regular manner' (Milne 1935a, 192). To describe the regular repetition of soil profiles on crest–hollow **topography**, which forms a convenient mapping unit, Milne chose the word 'catena' (from the Latin for a chain), which is equivalent to a toposequence (Jenny 1941). Key points in Milne's work were that all soils occurring along a hillslope relate to one another owing to the agency of geomorphic and soil processes, and that each soil zone supports a characteristic type of vegetation, so forming a vegetation catena (Milne 1935a, b).

The arrival of the catena concept prompted a spate of studies considering the relationships between soils and hillslopes. These studies led to refinements of the basic model, including the distinguishing of three zones (or complexes) along a catena, each associated with a broad topographic site: the eluvial zone, the colluvial zone, and the illuvial zone (Morison 1949). The eluvial zone is a high-level site that loses water and soluble and suspended matter. Material washed from it builds up the colluvial and illuvial zones. The colluvial zone occupies slope sites. It receives material from soils in the eluvial zone and loses some of it to the illuvial zone. The illuvial zone occupies low-level sites. In many cases, it has very mixed parentage, consisting of either a simple mosaic or else a mosaic of zoned patterns, depending upon the amount and nature of drainage. It has three distinguishing characteristics: it receives more water than the climatic normal site; it receives much dissolved and suspended matter; and water is lost from it by surface movement, by drainage, or by evaporation. During the 1960s, researchers started to take up Milne's and Morison's seminal ideas and investigate soil **evolution** in the context of landscapes. Several statements affirmed Morison's view that, because solution and water transport act selectively, the lateral concatenation of soils leads to the differentiation of soil materials. This means that the hill soils in a landscape may be thought of as A horizons, and the valley soils as B horizons (Blume and Schlichting 1965). These ideas, fostered by a consideration of the catena concept, have led to investigations of the lateral translocation of soil material along hillslopes. Subsequently, the burgeoning sophistication of hillslope hydrological investigations has prompted increasingly detailed and revealing examinations of slope soils (see Huggett 1995, 171–72).

Today, the catena is still an object of research and pedologists investigate relationships between soils and hillslopes. To do so, they employ two chief methods. First, they examine correlations between soil types or soil properties and hillslope elements (summit, shoulder, backslope, and so forth) or hillslope topographic variables (usually slope gradient, slope curvature, or aspect). Second, they establish causal connections between hillslope soils and soil processes and hillslope hydrology. This method may involve the field measurements of soil water and solute movements, the assessment of long-term translocation soils material using elemental ratios or index minerals, or mathematical modelling. Sundry studies correlate soil properties with particular topographic (hillslope) variables. The simplest cases measure a few soil properties and single topographic attributes at sites along toposequences (e.g. Swanson 1985), while more complicated cases investigate many soil properties and several topographic properties, and sometimes vegetation to boot (e.g. Chen *et al.* 1997). Soil–hillslope hydrology studies fall into two broad groups: those that measure the current movements of soil materials and water in the field (e.g. Litaor *et al.* 1998); and those that combine techniques for reconstructing the past movements of materials with an understanding of hillslope hydrology (e.g. Thompson and Bell 1998; Thompson *et al.* 1998).

Although research on soil and vegetation catenas undeniably has led to a better understanding of some soil and ecological processes, much work now focuses upon the three-dimensional **soil–landscape** of which a catena is but a part.

It is worth noting that Adrian E. Scheidegger (1986) enunciated a catena principle in geomorphology, which holds that all landscapes are a collection of catenas, and that each catena comprises an eluvial, colluvial, and eluvial zone. The eluvial zone lies at the top of the catena and consists of a flat summit and shoulder; the colluvial zone lies in the middle of the catena and consists of a backslope and foot-slope; the alluvial zone lying at the bottom of the catena is the toeslope and, like the summit, is relatively flat.

CHRONOSEQUENCE

Chronosequences are related sets of **ecosystems** (vegetation and soils) or landforms evolved under similar environmental conditions, but starting at different times. The spatial differences between the units translate into differences in time, the procedure sometimes being called space–time substitution or, using a term borrowed from physics, **ergodicity**. The translation hints at the way that ecosystems and land-forms evolve. Charles Darwin, for example, used the chronosequence method to test his ideas on coral-reef formation. He thought that barrier reefs, fringing reefs, and atolls occurring at different places represented different evolutionary stages of island development applicable to any subsiding volcanic peak in tropical waters. Likewise, William Morris Davis applied this evolutionary schema to landforms in different places and derived what he deemed was a time sequence of landform development – the **geographical cycle** – running from youth, through maturity, to senility.

Geomorphologists recognize topographic chronosequences. The best examples normally come from artificial landscapes, though there are some landscapes in which, by quirks of history, spatial differences are translatable into time sequences. Occasionally, field conditions lead to adjacent hillslopes being progressively cut off from the action of a fluvial or marine process at their bases. This has happened along a segment of the South Wales coast, in the British Isles, where cliffs are formed in Old Red Sandstone (Savigear 1952, 1956). Originally, the coast between Gilman Point and the Taff estuary was exposed to wave action. A sand spit started to grow. Wind-blown and marsh deposits accumulated between the spit and the original shoreline, causing the sea progressively to abandon the cliff base from west to east. The present cliffs are thus a topographic chronosequence: the cliffs furthest west have been subject to subaerial denudation without waves cutting their base the longest, while those to the east are progressively younger (Figure 6).

Soil chronosequences were once relatively rare, but the number of well-documented and reliably dated examples is mounting very fast. They are excellent indicators of the rate and direction of pedogenic change; they provide invaluable information for testing theories of **pedogenesis**; they are central to much soil–geomorphic research; and they may be used (with caution) to date a range of Quaternary land-forms, including stream terraces. Most soil chronosequences were in the past established using the comparative geographical technique, which involved placing soils of different age, and so from different locations, into an assumed time sequence that was thought represent successive stages of one or several pedogenic processes. The founda-tions of this method are very insecure. Nowadays, the construction of geographical soil chronosequences uses soils that have developed on surfaces of known age, and that either have persisted through to the present day or else have survived beneath a sedimentary cover. A common kind of soil chronosequence occurs in sequences of alluvial fans, alluvial terraces, coastal sand dunes, lava flows, and moraines, and also abandoned pasture, earthflows, mudflows, fire-cleared areas, land-slide scars, historically created polders, loess deposits, strandlines, and other datable landscape features. In soil chronosequences of this type, the soils started to develop at different times and continue developing

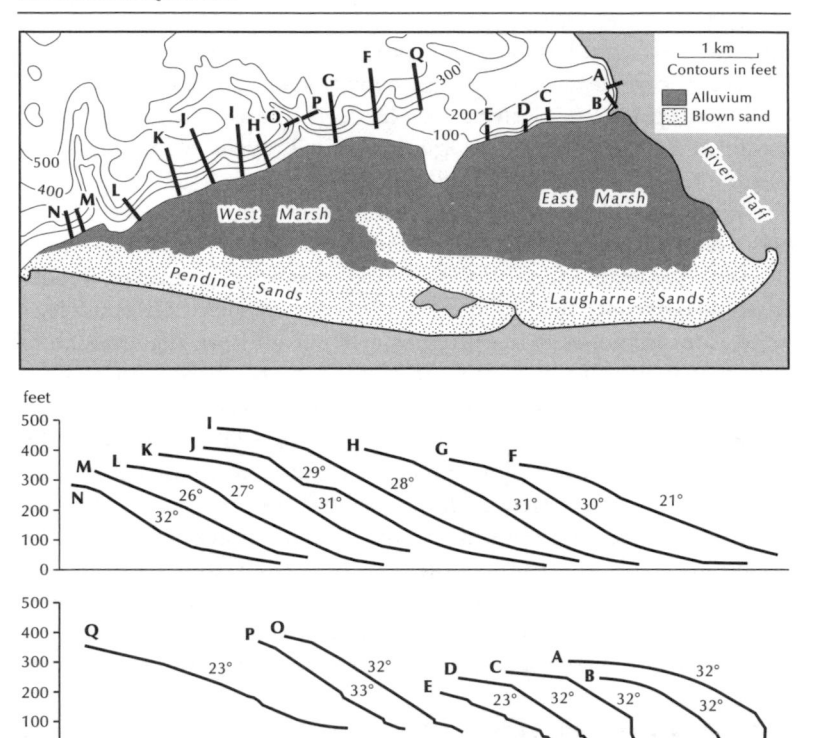

Figure 6 A topographic chronosequence in South Wales. (a) The coast between Gilman Point and the Taff estuary. The sand spit has grown progressively from west to east so that the cliffs to the west have been longest protected from wave action. (b) The general form of the hillslope profiles located in Figure 6a. Cliff profiles become progressively older in alphabetical order, A–N.

Source: After Huggett (1997a, 238), adapted from Savigear (1952, 1956)

at present. Soils formed in alluvial terraces provide an excellent example. Alluvial terraces generally form a staircase within a river valley, and in stepping down the staircase, soils become progressively younger so that the soils forming in each tread form a chronosequence, assuming that the soils on the younger terraces will evolve into the soils on the older terraces. A second kind of soil chronosequence occurs where the soils started developing at the same time but their development stopped at different times, owing to selective and successive burial by later events. An example would be soils evolved in a newly exposed glacial till that has been gradually covered by another deposit, so stopping pedogenesis and burying soils at different stages of their **evolution**. A third possibility is where soil formation starts and stops at different times with some overlap. Such chronosequences consist of a mixture of buried and relict soils produced by erosion and deposition. An example is a complex of buried and surface soils in lateral moraines, Bugaboo Glacier area, British Columbia, Canada (Karlstrom and Osborn 1992). A fourth type of soil chronosequence occurs where start and stop times differ and have no temporal overlap. These chronosequences lie in vertical sequences of soil–landscape units, such as those found between successive sedimentary units. In addition, the 'geological palaeosols' dating back to Precambrian times form a very disjointed chronosequence of sorts (e.g. Retallack 2001).

Vegetation chronosequences develop on landforms of differing age. Retreating glaciers expose new land ahead of the ice front that plants and animals can colonize. A classic study of primary **succession** was made in Glacier Bay National Park, south-east Alaska (Cooper 1923; Crocker and Major 1955). The glacier has retreated considerably since about 1750 (Figure 7). Proglacial settings also give insect community chronosequences (Hodkinson *et al.* 2004). On a smaller scale, lichens on stands of lodgepole pine (*Pinus contorta*) of different ages in Canada form a chronosequence (Coxson and Marsh 2001).

A word of caution is necessary here because not all spatial differences in ecosystems and landforms translate into temporal differences – factors other than time exert a strong influence on ecosystems and the form of the land surface, and ecosystems and landforms of the same age might differ through historical accidents. Moreover, it pays to be aware that, owing to **equifinality**, different sets of processes may produce the same ecosystem or landform.

Further reading: Birkeland 1990

CLIMATE CHANGE

Climate change is a change in average atmospheric conditions at a particular place or in a particular region. Scientists now understand the basic mechanisms of climate change in broad terms, although the relative importance of various factors over different time-spans is still uncertain. In the short term (over years, decades, and centuries), the atmosphere may change owing to external forcings (cosmic, geological, and anthropogenic) or to internal atmospheric dynamics. Cosmic forcing arises from changes in gravitational forces, by variations in

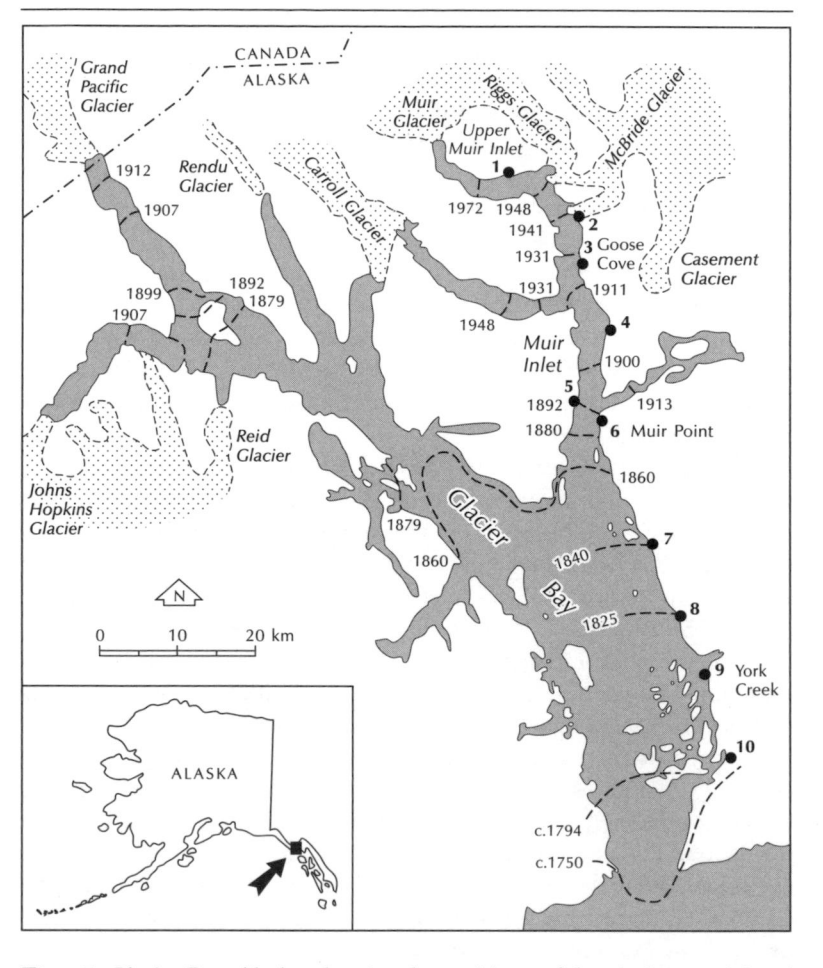

Figure 7 Glacier Bay, Alaska, showing the positions of the glacier termini and Fastie's (1995) study sites.

Source: Adapted from Crocker and Major (1955) and Fastie (1995)

electromagnetic and particulate radiation receipt from the Sun and from space, and by the impact of asteroids and comets. Gravitational stresses may emanate from three sources – the Solar System, our Galaxy, and other galaxies – but only interactions within the Solar System will force the atmosphere in the short term when the overall motions and alignments of the planets in the Solar System and by Earth–Moon motions modulates the delivery of **energy**. Geological forcing is, in the short term, caused by the injection of volcanic dust and gases into the stratosphere. Internal dynamics of the atmospheric system involve short-term, cyclical components. Thus, climate change can occur without the aid of external forcing. Sorting out signals from these potential sources of climate change is very hard. Humans cause climate change largely by interfering with **biogeochemical cycles**, especially the carbon cycle, and by changing land cover.

Paradoxically, change in modern climates is difficult to perceive – it commonly hides in the welter of direct measurements taken over short time periods; whereas change of past climates is unmistakable, owing to limited observations of proxy climatic indicators taken from rocks, sediments, and **soils**. Past climates (palaeoclimates) are no easier to define than modern climates. Within the historical period, various written accounts and instrumental records may provide clues to climatic conditions in the recent past. The stratigraphic record contains evidence of historical and older climates, but it does not always register conditions on a yearly basis and seldom supplies an annual calendar of palaeoclimatic events and states. Another difficulty is that not all past climates appear to have modern analogues. The disposition of land and sea, the position of high land, and other physical conditions were different in the past and, in some cases, produced climates that do not exist today – no-analogue climates. A prime example is the 'boreal grassland' climate that occupied a broad belt immediately south of the Northern Hemisphere ice sheets (see **no-analogue communities**).

For a variety of reasons, weather and climate change over timescales ranging from less than an hour, as in short-lived but severe meteorological phenomena, to over tens of millions of years, as in protracted phases of **global warming** and cooling (Table 3). In addition, they follow secular trends lasting aeons.

Further reading: Burroughs 2007; Cowie 2007; Fry 2007

Table 3 Major climatic variations, excluding secular trends

Variation	Period (years unless otherwise stated)	Nature of period	Nature of variation
Short-term variation			
Diurnal	1 day	Spike[a]	Daily cycle of climate
Weekly	3–7 days	Peak[b]	Synoptic disturbances, chiefly in middle latitudes
Annual	1	Spike	Annual cycle of climate
Quasi-biennial	~26 months	Peak	Wind shift (east phase to west phase) in tropical stratosphere (quasi-biennial oscillation – QBO)

continued overleaf

Table 3 Continued

Variation	Period (years unless otherwise stated)	Nature of period	Nature of variation
Quinquennial	2–7 years	Peak	Rapid switch of pressure distribution across the southern Pacific Ocean. Linked with changes in temperature in the same region, and together termed El Niño–Southern Oscillation (ENSO). Average period is 4–5 years; 2–7 years is the range
Undecennial	~11	Peak	Quasi-periodic variation corresponding to the sunspot cycle
Nonadecennial	~18.6	Peak	Quasi-periodic variation corresponding to the lunar nodal cycle
Octogintennial	~80	Peak	Quasi-periodic variation corresponding to the Gleissberg cycle
Bicentennial	~200	Peak	Quasi-periodic variation corresponding to the solar orbital cycle
Bimillennial	~2,000	Peak	Quasi-periodic variation of uncertain correspondence
Medium-term variation			
Precessional	~19,000	Spike	A periodic component of the Earth's precession cycle
Precessional	~23,000	Spike	A periodic component of the Earth's precession cycle
Tilt	~41,000	Spike	Chief periodic component of the Earth's tilt (obliquity) cycle
Short eccentricity	~100,000	Spike	A periodic component of the Earth's orbital eccentricity cycle
Orbital plane inclination	~100,000	Spike	A periodic component of Earth's orbit
Long eccentricity	~400,000	Spike	A periodic component of the Earth's orbital eccentricity cycle
Long-term variation			
Thirty megayear	~30,000,000	Peak	Quasi-periodic fluctuations perhaps corresponding to a tectonic cycle
Warm mode–cool mode	~150,000,000	Peak	Quasi-periodic fluctuations from warm to cool climatic modes, possibly connected to half a galactic year
Hothouse–icehouse	~300,000,000	Peak	Quasi-periodic fluctuations from hothouse to icehouse conditions

[a]Strictly periodic variation dictated by **astronomical forcing** cycles.
[b]Quasi-periodic variation with preferred timescales of occurrence.

Source: After Huggett (1997a, 109)

CLIMAX COMMUNITY

A climax community, or climatic climax community, is a community of plants and animals that, through the process of **succession**, has attained **equilibrium** with the local and regional environmental conditions. Frederic E. Clements (1916), by sharpening up ideas set down by Henry C. Cowles (1899), originated the idea of a single climatic climax, or monoclimax, associated with regional climate and representing the endpoint of succession. He believed each climatic type fosters a single climax-type, but recognized that other communities – he called them proclimax communities – exist that persist in states removed from the climatic climax for a particular area. He identified four such proclimax communities: subclimax (a long-lasting, penultimate stage of succession), disclimax or plagioclimax (caused by an environmental **disturbance**), preclimax (caused by drier or warmer conditions than the regional norm), and postclimax (caused by cooler or wetter conditions than the regional norm). However, he thought these communities were unstable, because by definition climax vegetation is best adapted to the climate of a given area (see Eliot 2007). Arthur Tansley extended Clement's idea with the polyclimax in which several different climax communities could exist in an area with the same regional climate owing to differences in soil moisture, nutrient levels, fire frequency, and so on (e.g. Tansley 1939). The climax-pattern hypothesis (Whittaker 1953) was a variation on the polyclimax theme and saw natural communities adapted to all environmental factors, but with a continuity of climax types that grade into one another along environmental gradients, rather than forming discrete communities that change through very sharp **ecotones**.

A recent rereading of Clements revealed several misconceptions about what Clements actually said (Eliot 2007). Clements never argued that climax communities must always occur, or that climate is the chief determinant of vegetation, or that the different species in an ecological community form a tightly integrated physiological unit, or that plant communities have sharp boundaries in time or space. Instead, he used the idea of a climax community as a conceptual starting point for describing the vegetation in a given area. There are good reasons to believe that the species best adapted to some conditions might appear there, when those conditions occur. Nevertheless, much of Clements's work was committed to characterizing what happens when the ideal conditions do not occur. In those circumstances, vegetation other than the ideal climax will often occur instead. However, Clements opined that those different kinds of vegetation were describable as deviations from the climax ideal. In consequence, he developed a very large vocabulary of theoretical terms describing the possible causes of vegetation, and various non-climax states vegetation adopts as a result.

Owing to difficulties in applying Clements's terms, to the demonstration of vegetation gradients between communities, and to the promulgation of other and seemingly superior theories that indicated the general instability and impermanence of communities and the multidirectional character of some successional sequences (see **community change**), support for the climatic climax community concept waned (Tobey 1981). Nonetheless, although ecologists dumped the climatic climax theory, its parlance enjoyed a new airing during the 1990s in theoretical ecology (e.g. Roughgarden *et al.* 1989). Moreover, a number of authors and Nature enthusiasts still use the term 'climax' in a watered-down form to describe mature or old-growth communities.

COMMUNITY CHANGE

Clements's view of vegetation **succession** and **climatic climax communities** dominated notions of community change for much of the twentieth century. However, always lurking in the background were the views of Henry Allan Gleason (1926) and Alexander Stuart Watt (1924) who saw communities and **ecosystems** as not being in **equilibrium**, at least not in Clements's sense of stable climax communities. The disequilibrium view rose to stardom in the early 1970s, when some ecologists dared to suggest that **succession** leads nowhere in particular so that there are no long-lasting climatic climaxes (e.g. Drury and Nisbet 1973). Instead, each species 'does its own thing', communities are ever-changing and temporary alliances of individuals, and succession runs in several directions (it is multidirectional, not unidirectional). This disequilibrium view emphasizes the individualistic behaviour of species and the evolutionary nature of communities. It stresses imbalance, disharmony, **disturbance**, and unpredictability in Nature. In addition, it focuses on the geography of ecosystems – landscape patches, corridors, and matrixes replace the climax formation and ecosystem, and landscape mosaics replace the assumed homogeneous climax and ecosystem.

The individualistic comings and goings of species influence community change in a profound way. Communities change because new species arrive and old species are lost. New species appear in **speciation** events and through immigration. Old species vanish through local **extinction** (extirpation) and through emigration. Some species increase in abundance and others decrease in abundance, thus tipping the competitive balance within a community. Each species has its own propensity for **dispersal**, invasion, and **population** expansion. Community assembly is an unceasing process of species arrivals, persistence, increase, decrease, and extinctions played out in an individualistic way. Multidirectional succession and community impermanence furnish evidence that communities assemble (and disassemble) in this manner. A result of individualistic community assembly is that succession may continue along many pathways, and is not necessarily fenced into a single predetermined path – it is multidirectional and not unidirectional. Some field studies of historical community change – for instance, Glacier Bay, Alaska (Fastie 1995), Krakatau Islands, Indonesia (Whittaker *et al.* 1989, 1992; Whittaker and Jones 1994), and Mauna Loa, Hawaii (Kitayama *et al.* 1995) – support this powerful notion of multidirectional succession.

If each species 'does its own thing', then communities, both local ones and biomes, should come and go in answer to **environmental changes**. This argument leads to a momentous conclusion: there is nothing special about present-day communities and biomes. However, contrary to this bold assertion, insect species and communities have shown remarkable constancy in the face of Quaternary climatic fluctuations (Coope 1994). Research on communities with no modern analogues and on computer-generated communities indicates the impermanence of communities. **No–analogue communities** are dealt with elsewhere, but in brief, some modern communities and biomes are similar to past ones, but most have no exact fossil counterparts; contrariwise, many fossil communities and biomes have no precise modern analogues. An example comes from the Missouri–Arkansas border region, USA, where from 13,000 to 8,000 years ago, the eastern hornbeam (*Ostyra virginiana*) and the American hornbeam (*Carpinus caroliniana*) were significant plant community components (Delcourt and Delcourt 1994). These communities, which lay between the Appalachian Mountains and the Ozark Highlands, bore little resemblance to any modern communities in eastern North America, and they seem to have evolved in a climate characterized by heightened seasonality and springtime peaks in solar radiation. Some mathematical models also point to community impermanence and the individualistic action of component species. In one model, experiments started with a 125-species pool of plants, herbivores, carnivores, and omnivores (Drake 1990). The model selected one species at a time to join an assembling community, with second chances permitted for first-time failed entrants. An extremely persistent community emerged comprising about fifteen species. When rerun with the same species pool, the model again produced an extremely persistent community, but this time with different component species than in the first community. There was nothing special about the species in the communities: most species could become a member of a persistent community under the right circumstances; the actual species present depended on happenstance. The dynamics of the persistent communities was very special: it was impossible to reassemble a persistent community of fifteen species from scratch using only those fifteen species. This finding suggests that communities are not artificially constructible from a particular set of species – they have to evolve and to create themselves out of a large number of possible species interactions.

COMPLEXITY

Complexity is the stunning notion that simple rules govern all complex **systems**. It is a way of describing complicated, irregular patterns that appear random, such as turbulent flow in streams and the interactions between predator and prey **populations**. Much chaotic complexity in environmental systems underlies a larger-scale order, and overlies smaller-scale, more orderly and understandable components. Chaotic turbulent flow is part of a larger-scale order seen in the predictable rate and direction of mean streamflow; it is also the result of a huge number of well-understood individual particle trajectories describable by the basic laws of physics. Predator–prey interrelationships are part of a larger-scale order seen in total population numbers, and at the same time the outcome of myriad interactions between individual predators and their prey. Complexity in environmental systems is thus often part of a hierarchy of interrelated structures and processes. Similarly, simple patterns in environmental systems, such as beach cusps, commonly arise from complex underlying dynamics; at the same time, they are but a part of broader-scale complex patterns. Beach cusps result from complex nonlinear interactions between beaches and waves or the complicated formation of edge waves (waves trapped at the shoreline by refraction); at the same time, they are a part of irregular coastline geometry.

Complexity has its roots in classical open systems research, which characteristically dealt with linear relationships in systems near **equilibrium**. A fresh direction in thought and a deeper understanding came with the discovery of deterministic chaos by Edward Lorenz in the early 1960s (Lorenz 1963a, b). Technically speaking, this was a rediscovery as Henri Poincaré had dealt with similar issues in nonlinear mechanics (e.g. Poincaré 1881–86). However, deterministic chaos did not make a grand entrance into the scientific mindset until the 1960s. The key change was the recognition of nonlinear relationships in systems. In environmental systems, nonlinearity means that system outputs (or responses) are not proportional to systems inputs (or forcings) across the full gamut of inputs (cf. Phillips 2006a).

Nonlinear relationships produce rich and complex dynamics in systems far removed from equilibrium, which display periodic and chaotic behaviour. The most surprising feature of such systems is the generation of 'order out of chaos', with systems states unexpectedly moving to higher levels of organization under the driving power of internal entropy production and entropy dissipation. Systems of this kind, which dissipate **energy** in maintaining order in states removed from equilibrium, are dissipative systems. It is perhaps useful to distinguish 'simple' evolving systems, such as planets, stars, and galaxies, from complex adaptive systems that learn or evolve by utilizing acquired information, as when a child learns his or her native language, a strain of a bacterium becomes resistant to an antibiotic, and the scientific community tests new theories (Gell-Mann 1994).

In dissipative systems, non-equilibrium is the source of order, with spontaneous fluctuations growing into macroscopic patterns. The Bénard convective cell is an instructive example (Prigogine 1980, 88). Imagine a horizontal layer of fluid at rest between two parallel planes. Warm the bottom plane and hold it a higher temperature than the top plane. When large enough, the temperature gradient between the two planes causes the state of rest to destabilize and **convection** begins. Entropy production increases because the convection is a new mechanism for heat transport. In more detail, while the fluid is at rest and below the **threshold** temperature gradient, small convection currents appear as fluctuations from the average state but they are damped and disappear. Above the critical temperature gradient, some of the fluctuations amplify to produce a macroscopic current. In effect, the fluctuations trigger an instability that the system accommodates by reorganizing itself. The macroscopic convective cell stabilizes by exchange of energy with the system's **environment**. The **general circulation of the atmosphere** works on the same principle.

The theory of complex dynamics predicts a new order of order, an order arising out of, and poised perilously at the edge of, chaos. It is a fractal order that evolves to form a hierarchy of spatial systems whose properties are holistic and irreducible to the laws of physics and chemistry. Geomorphic examples are flat or irregular beds of sand on streambeds or in deserts that self-organize themselves into regularly spaced forms – ripples and dunes – and are rather similar in size and shape (e.g. Baas 2002). Conversely, some systems display the opposite tendency – that of non-self-organization – as when relief reduces to a plain. A central implication of chaotic dynamics for the natural world is that all Nature may contain fundamentally erratic, discontinuous, and inherently unpredictable elements. Nonetheless, nonlinear Nature is not all complex and chaotic. Phillips (2006a) sagely noted that 'Non-linear systems are not all, or always, complex, and even those which can be chaotic are not chaotic under all circumstances. Conversely, complexity can arise due to factors other than nonlinear dynamics'.

One of the most remarkable features of complex systems is their behaviour. Complex systems are sensitive to initial conditions, a notion popularized as the Butterfly Effect (in which a butterfly fluttering its wings in England causes a hurricane in Australia – I wonder if the reverse is this true, could a hurricane in Australia prompt a butterfly to flap its wings in England?). They obey simple deterministic laws, but their behaviour is irregular. Indeed, it may be so irregular that it looks random. However, chaotic behaviour is not random; it is a cryptic, random-like pattern generated by simple deterministic laws. So, contrary to the traditional view that simple causes must produce simple effects (and the implied corollary that complex effects must have complex causes), chaos theory predicts that simple causes can create complex effects. Because of this, knowledge of the simple deterministic rules governing the behaviour of a complex system does not guarantee success in predicting the system's future behaviour. However, it does mean that, for instance, landscape models do not need to become increasingly complex to give useful predictions (Favis-Mortlock and de Boer 2003). Significantly, a system displaying chaotic behaviour through time usually displays spatial chaos, too. Thus, a landscape that starts with a few small perturbations here and there, if subject to chaotic **evolution**, displays increasing spatial variability as the perturbations grow (Phillips 1999b, 20).

Culling (1987, 1988) recognized the potential importance of nonlinear dynamics for geomorphic thinking (see also Huggett 1988). Phillips is surely the most dogged and industrious proponent of nonlinear dynamics in Earth surface systems. From his studies, he drew up eleven principles of Earth surface systems, which illustrate the potency of the non-equilibrium paradigm (Phillips 1999b; see also Huggett 2007a, 405). More recently, he has stressed the importance of confronting nonlinear complexity by 'problematizing nonlinear dynamics from within a geomorphological context', rather than applying analytical techniques derived from mathematics, statistics, physics, and other disciplines that use experimental laboratory techniques and numerical models (Phillips 2006a). To this end, and rooting his arguments in field-based studies, he discussed methods for detecting chaos in geomorphic systems (and other environmental systems), explored the idea of unstable and non–equilibrium systems versus stable systems achieving a new equilibrium following a change in boundary conditions, and shed a new light on the question of space and time scales. Convergence versus divergence of a suitable system metric (elevation or regolith thickness for instance) is a hugely significant indicator of stability behaviour. In landscape evolution, convergence associates with downwasting and a reduction of relief, whilst divergence relates to dissection and an increase of relief. More fundamentally, convergence and divergence underpin developmental, 'equilibrium' conceptual frameworks, with a monotonic move to a unique endpoint (peneplain or other steady-state landform), as well as evolutionary, 'non–equilibrium' frameworks that engender historical happenstance, multiple potential pathways and end-states, and unstable states. The distinction between instability and new equilibria is critical to understanding the dynamics of actual geomorphic systems, and for a given scale of observation or investigation, it separates two conditions. On the one hand is a new steady-state equilibrium governed by stable equilibrium dynamics that develops after a change in boundary conditions or in external forcings. On the other hand is a persistence of the disproportionate impacts of small disturbances associated with dynamic instability in a non–equilibrium system (or a system governed by unstable equilibrium dynamics) (Phillips 2006a). The distinction is critical because the establishment of a new steady-state equilibrium implies a consistent and predictable response throughout the system, predictable in the sense that the same changes in boundary conditions affecting the same system at a different place of time would produce the same outcome. In contrast, a dynamically unstable system possesses variable modes of system adjustment and inconsistent response, with different outcomes possible for identical or similar changes or disturbances. Several indicators potentially allow the identification of newly stable equilibria and dynamically unstable system states in field situations.

Related to complexity are the ideas of fractals and self-organized criticality. Fractal landscapes, for instance, display self-similar patterns repeated across a range of scales (Xu *et al.* 1993). A small section of coastline may be self-similar to a much larger piece of coastline, of which it is part. Drainage networks, sedimentary layers, and joint systems in rocks possess fractal patterns (Hergarten and Neugebauer 2001). Self-organized criticality is a theory that systems composed of myriad elements will evolve to a critical state, and that once in this state, then tiny perturbations may lead to chain reactions that may affect the entire system. The classic example is a pile of sand. Adding grains one by one to a sandbox causes a pile to start growing, the sides of which become increasingly steep. In time, the slope angle becomes critical: one more grain added to the pile triggers an avalanche that fills up empty areas in the sandbox. After adding sufficient grains, the sandbox overflows. When, on average, the number of sand grains entering the pile equals the number of grains leaving the pile, the sand pile has self-organized into a critical state. Landslides, drainage networks, and the **magnitude and frequency** relations of earthquakes display self-organized criticality.

Further reading: Bradbury *et al.* 1996; Gribbin 2004; Richards 2002; Stewart 1997

CONTINENTAL DRIFT

This is the movement of the Earth's landmasses relative to each other. Frank Bursley Taylor proposed the concept in 1908 (see Taylor 1910), arguing that the continents separated during the Tertiary period. Alfred Lothar Wegener was the first to use the term continental drift (*die Verschiebung der Kontinente* in his native German) and to publish formally the hypothesis that the continents had somehow drifted apart (Wegener 1912, 1915, 1966). However, he was unable to provide a convincing explanation for the physical processes that might have caused this drift. The scientific community considered unrealistic his suggestion that the centrifugal pseudo-force of the Earth's rotation pulled the continents asunder. Consequently, the hypothesis of continental drift remained controversial, although it received robust support from British geologist Arthur Holmes and South African geologist Alexander Du Toit.

The idea of continental drift did not become widely accepted even as a theory until the late 1950s. By the 1960s, geological research conducted by Robert S. Dietz, Bruce Heezen, and Harry Hess, along with a rekindling of the theory that includes a driving mechanism by J. Tuzo Wilson, led to its widespread acceptance among geologists, and it became part of the more inclusive theory of **plate tectonics**.

Wegener, the man who started it all, was perhaps one of the most important scientific innovators of the twentieth century (Hallam 1973, 114).

CONVECTION

Convection, in broad terms, is the mass movement of currents within fluids – liquids, gases, and rheids (solid materials that deform by viscous flow). It is one of the major modes of heat and mass transfer through the atmosphere and oceans, and occurs within the Earth's interior (see **plate tectonics**). Convection arises under two circumstances. First, it occurs when the moving fluid flows over an uneven boundary surface, which generates irregular eddies (forced convection or mechanical turbulence). Second, it occurs when the fluid is heated from below to produce less dense and therefore more buoyant parcels that rise as upward flowing currents (thermals in the air, plumes in the mantle) that adjacent downward flowing currents replenish (free convection) to form convective cells. Such convection involves vertical movements, which in meteorological **systems** range in size from clouds and thunderstorms to the **general circulation of the atmosphere**.

In fluids, convective heat and mass transfer take place by diffusion, which is the random (or Brownian) motion of individual particles in the fluid, and by **advection**, in which the larger-scale motion of currents in the fluid transports matter or heat. When speaking of heat and mass transfer, the term 'convection' refers to the sum of advective and diffusive transfer. The term convection also applies to the special case in which the advected (carried) substance is heat. In this case, the heat itself often causes the fluid motion (as in a convective cell), while also being transported by it, and the problem of heat transport (and the related transport of other substances in the fluid due to it) can be complicated.

Further reading: Davies 1999; Emanuel 1994

CYCLICITY/PERIODICITY

Cycles are things or events that recur at regular intervals. Examples are diurnal cycles, seasonal cycles, and annual cycles. A period is the time between the same points on a cycle – a day, a year, or whatever. Cyclicity and periodicity are hugely important in all branches of physical geography, mainly because a vast number of biological, climatological, ecological, geological, geomorphological, hydrological, meteorological, and pedological phenomena display cyclical behaviour. This fact was eloquently expressed by Joseph Barrell:

> Nature vibrates with rhythms, climatic and diastrophic, those finding stratigraphic expression ranging in period from the rapid oscillation of surface waters, recorded in ripple-mark, to those long-deferred stirrings of the deep imprisoned titans which have divided earth history into periods and eras.
>
> (Barrell 1917, 746)

A large number of well-established external and internal cycles potentially affect Earth **systems**, although the detection of some of them in the environmental record is uncertain. The chief external cycles relate to solar activity, lunar cycles, planetary cycles, and galactic and intergalactic cycles. Solar cycles have several components ranging from days to centuries that have an impact on climate and other environmental systems (see **solar forcing**). Lunar cycles have strong daily, fortnightly, and 18.6-year periods. These affect environmental systems, with the 18.6-year lunar nodal or Metonic cycle (after Meton, a Greek believed to have discovered it) detectable in flood records, including those for the River Nile, in drought records (e.g. Currie 1984), and in oscillations in Arctic climate (Yndestad 2006). Planetary cycles associated with orbital forcing have significant components in the range 20,000 to 400,000 years and, under the right conditions, have a large impact climate, driving glacial and interglacial switches (see **astronomical forcing**).

The internal cycles are geological in origin, although these may partly lock into external cycles, and they are mostly very long-term cycles. A recent statistical analysis on high-resolution records of calcareous plankton diversity, global sea level, ratios of marine isotopes (oxygen, carbon, and strontium), large igneous province eruptions, and dated impact craters over the last 230 million years pointed to astrophysical or geophysical pacemakers of diversity change with periods in the range 25 to 33 million years (Prokoph *et al.* 2004). The results suggested that since the early Mesozoic debut of plankton, long-term cyclical changes in global environmental conditions and periodic large volcanic and impact events have modulated their diversity (Prokoph *et al.* 2004). Another study revealed a strong 62-million year cycle in diversity (Rohde and Muller 2005).

DESERTIFICATION

In 1949, Auguste Aubréville, a French forester, noticed that the Sahara Desert was expanding into surrounding savannahs and coined the term desertification to describe the process. Not until the 1970s did the term come to wide notice when a ruinous drought in the Sahel region of Africa led to the United Nations Conference on Desertification (UNCOD) in 1977, which showed that the process was probably occurring in all the world's drylands. The topic has since generated a huge literature, a legion of definitions, a collection of world maps, and much controversy.

With over a hundred definitions proposed, the meaning of desertification has proved difficult to pin down. In essence, the process of desertification degrades land in arid, semi-arid, and dry subhumid areas, reducing the land's capacity to accept, store, and recycle water, **energy**, and nutrients. However, it is a process with climatic, ecological, and human dimensions, at the heart of which lie links between biophysical and socio-economic processes and their effects on human welfare (Geist 2005, 2). As a concept, desertification relates to several other key ideas, including **carrying capacity**, land capability, and **sustainability**.

The primary causes of desertification are climatic variations, ecological change, and socio-economic factors, although the details of cause and effect are intricate. At root, desertification occurs because dryland **ecosystems** are vulnerable to certain climatic changes and overexploitation and unsuitable land use – drought, poverty, political instability, deforestation, overgrazing by livestock, overcultivation, and bad irrigation practices can all weaken the land's fertility and allow degradation to take hold. **Soil** compaction and crusting, quarrying, and desert warfare may also be causative factors in some cases. Whatever its causes, desertification directly affects over 250 million people, and puts at risk some one billion people in over a hundred countries, which is why it has generated so much research, to which physical geographers have made valuable contributions.

Further reading: Geeson *et al.* 2002; Geist 2005; Laity 2008; Middleton and Thomas 1997; Millennium Ecosystem Assessment 2005

DIRECTIONALISM

Charles Lyell was adamant that the world was in a steady state, displaying no overall direction in its history. It did change in the long term, but only about a mean condition. So precious to Lyell was his uniformity of state that, for most of his career he maintained that, since the Creation, life displayed no overall direction, to the extent that he believed that one day a fossil Silurian rat would turn up. Encapsulating Lyell's arguments, Stephen Jay Gould wrote:

> Land and sea would change places as the products of continents slowly eroded to fill up oceans, but land and sea would always exist in roughly constant amounts. Species would die and new ones would arise, but the mean complexity of life would not alter and its basic designs, created at the beginning, would endure to the end of time.
>
> (Gould 1984, 9)

Eventually, the burden of proof for directionality in the geosphere and biosphere became so overpowering that Lyell conceded directional change in life history.

Since the nineteenth century, more and more evidence of directional change has amassed so that no scientist would now attempt to uphold Lyell's steady-state interpretation of Earth history. The evolving states of the atmosphere and the evolving states of sedimentary rocks bear out directionality in the geosphere. In the biosphere, an increase in the complexity of life, an increase in the size and multicellularity of life, and an increase in the diversity of life all bespeak directionality. Admittedly, the burgeoning diversity of life has not followed a smooth, monotonic progression. The fossil record seems to show periods of relatively stable species composition broken by short periods of species change. Nonetheless, the scientific evidence for directional change in Earth history seems incontrovertible.

Further reading: Huggett 1997b

DISPERSAL

All organisms can, to varying degrees, move from their 'birthplaces' to new locations. Terrestrial mammals can walk, run, dig, climb, swim, or fly to new areas. The adults of higher plants and some aquatic animals are sessile (rooted to one spot), but are capable of roving large distances in their early stages of development. Dispersal occurs when organisms move to, and attempt to colonize, areas outside their existing range. It excludes seasonal migrations and population irruptions, as in the desert locust (*Schistocerca gregaria*) that swarms northwards from its central African core. The stage in the life cycle of an organism that does the dispersing is a propagule. In plants and fungi, a propagule is the structure that serves to reproduce the species – seed, spore, stem, or root cutting. In animals, a propagule is the smallest number of individuals of a species able to colonize a new area. Species may disperse by active movement (digging, flying, walking, or swimming), or by passive carriage. Physical agencies (wind, water, landmasses) or biological agencies (other organisms, including humans) bring about passive dispersal. These various modes of transport have technical names – anemochore for wind dispersal, thalassochore for sea dispersal, hydrochore for water dispersal, anemohydrochore for a mixture of wind and water dispersal, and biochore for hitching a ride on other organisms. Dispersal abilities of different groups of organisms vary enormously. Bats and land birds, insects and spiders, and land molluscs form the 'premier league' of transoceanic dispersers. Lizards, tortoises, and rodents come next, followed by small carnivores. The poorest dispersers are large mammals and freshwater fish.

Organisms disperse in at least three different ways: jump dispersal, diffusion, and secular migration. Jump dispersal is the rapid transit of individual organisms across large distances, often across inhospitable terrain, the jump taking less time than the life-span of the individual involved. This appears to be the method by which green iguanas (*Iguana iguana*) colonized Anguilla in the Caribbean (Censky *et al.* 1998). On 4 October 1995, at least 15 iguanas appeared on beaches on Anguilla, having arrived on logs and uprooted trees during a hurricane. Diffusion is the relatively gradual spread or slow penetration of populations across hospitable terrain over many generations. A case is the nine-banded armadillo (*Dasypus novemcinctus*) that has spread, and is still spreading, from Mexico to the south-western USA (Taulman and Robbins 1996). Secular migration is the spread or shift of a species that takes place very slowly, indeed so slowly that the species undergoes evolutionary change while it is taking place – by the time the population arrives in a new region, it will differ from the ancestral population in the source area (Mason 1954). South American members of the family Camelidae (the camel family), including the llama (*Lama glama*), are examples. They are all descended from now extinct North American ancestors that underwent a secular migration during Pliocene times over the then newly created Isthmus of Panama. The dispersal of horses from their North American homeland to Europe (and subsequent **extinction** in North America) and the diversification and spread of the Angiosperms (flowering plants) are other examples.

The process of dispersal is irrefutable, but biogeographers debate its significance in explaining the history of organisms. Charles Darwin and Alfred Russel Wallace laid the foundations of modern dispersal biogeography during the second half of the nineteenth century. They argued that species originated in a particular place (their centre of origin) and then spread elsewhere, crossing such barriers as mountains and seas in doing so. In the twentieth century, such illustrious researchers as Ernst Mayr, George Gaylord Simpson, Philip J. Darlington refined this classic centre-of-origin dispersal model. The ascendancy of **vicariance** as an explanation for historical biogeographical patterns during the late 1970s knocked dispersal biogeography from its long-standing position as the ruling theory. However, dispersal biogeography is making a vigorous comeback (McClone 2005; Queiroz 2005; Cowie and Holland 2006).

Further reading: Brown *et al.* 2005

DISTURBANCE

Disturbance has been a popular theme in ecological studies since the 1970s (e.g. Barrett *et al.* 1976). Disturbance is any event that disrupts the day-to-day running of an **ecosystem**. Disturbing agencies may be physical or biological. Strong wind, fire, flood, landslides, and lightning cause physical disturbance. Pests, pathogens, and the activities of animals, plants, **invasive species**, and humans cause biological or biotic disturbance. The effects of these disturbing agencies on ecosystems can be dramatic and alter the nature of communities. Pathogens, for example, are forceful disrupters of ecosystems, witness the efficacy of Dutch elm disease in England and the chestnut blight in the Appalachian region eastern United States.

Some disturbances act essentially randomly within a landscape to produce disturbance patches. Strong winds commonly behave in this way. The patches produced by random disturbance can be extensive: the patches of eroded **soil** created by grizzly bears (*Ursus arctos horribilis*) excavating dens, digging for food, and trampling well-established trails are a case in point (Butler 1992). Second, some disturbances, such as fire, pests, and pathogens, tend to start at a point within a landscape and then spread to other parts. In both cases, the disturbances operate in a heterogeneous manner because some sites within landscapes will be more susceptible to disturbing agencies than other sites.

It is common for a natural disturbance to allow new species to replace some of those in the pre-disturbance community, owing to changes in abiotic conditions and lower levels of competition. In consequence, the effects of disturbance may last for some time.

Nonetheless, in the absence of further disturbance, the ecosystem may revert to its original state through the process of **succession**, when it may then again be vulnerable to disturbance. In these circumstances, disturbance tends to drive a cycle of change. Pine forests and mountain pine beetle (*Dendroctonus ponderosae*) outbreaks in western North America provide a good example of such a cycle (Mock *et al.* 2007). The mountain pine beetle limits such pine trees as lodgepole pine (*Pinus contorta*) in forests of western North America. The beetles exist in endemic and epidemic phases. During epidemic phases, swarms of beetles kill large numbers of old pines, so creating openings in the forest for new vegetation. The beetles do not affect spruce, fir, and younger pines, all of which thrive in the canopy openings. Eventually, pines re-establish themselves in the canopy, replacing those lost. This cycle of death and regrowth generates a temporal mosaic of pines in the forest. Similar cycles stem from forest fires and wind-throw. Disturbing agencies often work in tandem. In forested landscapes of the south-eastern United States, individual pine trees are disturbed by lightning strikes. Once struck, pine trees are susceptible to colonization by bark beetles whose populations can expand to epidemic proportions and create forest patches in which gap-phase succession is initiated; thus the bark beetle appears to magnify the original disturbance by lightning (Rykiel *et al.* 1988).

Some species thrive on disturbance. In forests, shade-intolerant species rapidly fill gaps created by fire, wind-throw, or humans. In boreal forests, jack pine (*Pinus banksiana*) and some other pine species are adapted to crown fires – they have specialized serotinous cones that only open and disperse seeds with sufficient heat generated by fire (Schwilk and Ackerly 2001). In some cases, entire communities adapt to the disturbance regime. For instance, individual fires destroy and disturb susceptible communities. If the fire regime persists, the vegetation will become adapted to fire. At that point, the long-term survival of the community requires that fire dispense with invaders that are not fire-adapted. Therefore, the fire-adapted community has integrated fire as part of the system. Paradoxically, fire disturbance destroys the biomass of individuals but sustains the community of which the individuals are part. In the 373-ha Swartboskloof catchment, near Stellenbosch, Cape Province, South Africa, the vegetation is dominantly mesic mountain fynbos, a Mediterranean-type community, with forests in wetter sites around perennial streams and on boulder screes below cliffs. Major fires occurred in 1927, 1942, and 1958. These fires burnt the entire catchment. Smaller fires burnt parts of the catchment in 1936, 1973, and 1977. All the fires were accidental, save the prescribed burn in 1977. Research shows that the fires affected single species, communities, and the entire catchment ecosystem (Richardson and van Wilgen 1992).

Disturbance affects all scales of landscape (Figure 8). At small scales (1 to 500 years and 1 m² to 1 km²), wildfire, wind damage, clear-cut, flood, and earthquake are the dominant causes of disturbance events. Vegetation units at this **scale** range from individual plants and forest stands, and landscapes range from sample plots to first-order **drainage basins**. Local disturbances lead to patch dynamics within individual vegetation patches. Disturbance events over medium scales (500 to 10,000 years and 1 km² to 10,000 km²), encompass interglacial stages and landscapes ranging from second-order drainage basins to smaller mountain ranges. On the lower end of this scale, a prevailing disturbance regime, such as pathogen outbreaks and frequent fires, influences patch dynamics over a landscape mosaic. In the upper range of

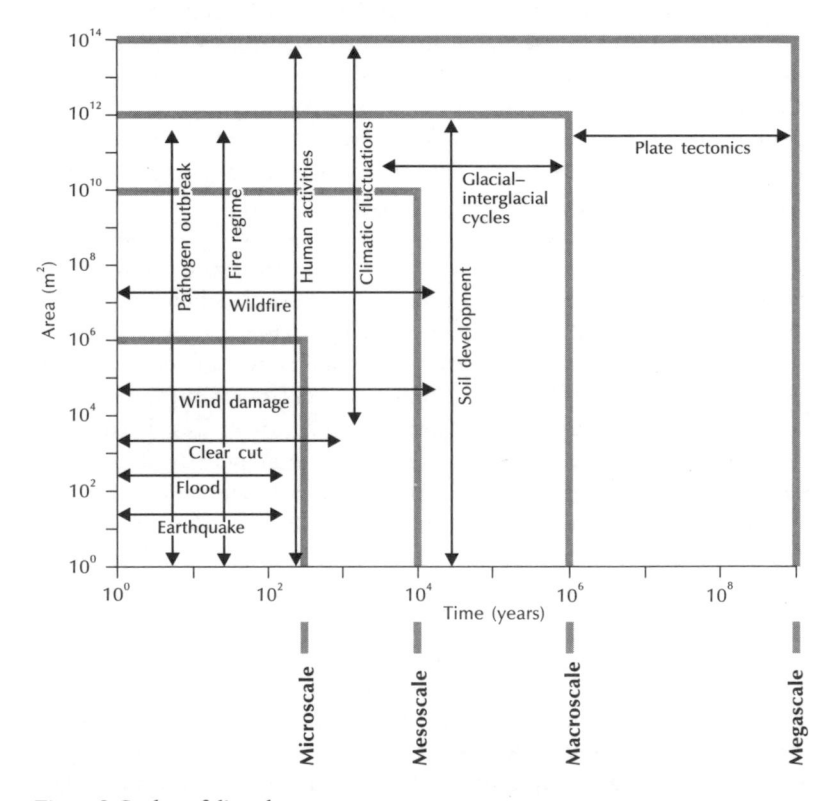

Figure 8 Scales of disturbance.

Source: Adapted from Delcourt and Delcourt (1988)

the scale, the prevailing disturbance regimes may themselves change so causing changes within patches and between patches that in turn alter the landscape mosaic. Disturbances at large scales (10,000 to 1,000,000 years and 10,000 km^2 to 1,000,000 km^2), span one to several glacial–interglacial cycles and affect landscapes ranging in size from physiographic provinces to subcontinents. At this scale, regional and global **environmental changes** cause changes in prevailing disturbance regimes. **Plate tectonics**, largely through its influence of global climates and sea level, causes disturbances at the largest scale, defined as more than a million years and areas of continental size and larger. **Bombardment** and volcanic activity also disturb ecosystems. Bombardment disrupts ecosystems within seconds, disturbing small or very large areas depending on the size of the impactor. Volcanic activity, like bombardment, acts very quickly and may disturb the entire ecosphere, though local disturbances are far more common. Repeated disturbance during a bout of volcanism is, arguably, more likely to cause severe stress in the ecosphere than is a single volcanic explosion.

Further reading: Dale *et al.* 2001; Johnson and Miyanishi 2007; Whelan 2008

DRAINAGE BASIN

A drainage basin is an area of land in which water from rain or snow-melt drains downhill into a river, lake, or sea. It includes the land surfaces from which the water drains, and the rivers that collect and convey the water. A drainage basin funnels all the water within its area into a waterway or water body. A water divide or a watershed (usually a ridge, hill, or mountain) separates adjacent drainage basins. In North America, the term watershed usually refers to the drainage basin itself. Catchment, catchment area, catchment basin, drainage area, river basin, and water basin are other equivalent terms for a drainage basin. Some 82 per cent of drainage basins drain into an ocean (exorheic basins); the remaining 18 per cent are endorheic basins and drain to inland lakes or seas. Much of interior Asia, for instance, drains into the Caspian and Aral Seas.

The concept of a drainage basin has relevance in many disciplines, including hydrology, geomorphology, and ecology. In hydrology, the drainage basin, bounded by its watershed and subject to surface and subsurface drainage to the ocean or interior lakes, is a logical unit of focus for studying the movement of water within the land leg of water cycle. The framework of the drainage basin allows the drawing up of water balances, the assessing of water resources, and the prediction of extreme events such as floods and droughts. As drainage basins are coherent hydrological units, they provide rational structures for managing water resources. In Minnesota, USA, governmental bodies that manage water resources are called watershed districts; in New Zealand, they are catchment boards; and comparable community groups based in Ontario, Canada, are called conservation authorities. In geomorphology, drainage basins are restricted, handy, and usually clearly defined and unambiguous topographic units, available in a nested hierarchy of sizes; they are also open physical **systems** in terms of inputs of solar radiation and precipitation, throughputs of surface and subsurface water flows, and outputs of discharge, evaporation, and long-wave radiation (Chorley 1969). In ecology, drainage basins are important elements of **ecosystems**. Water flowing over the ground and along rivers can pick up nutrients, sediment, and (in some places) pollutants, which the water carries towards the basin outlet. On route and in the receiving water body (lake or ocean), the nutrient, sediment, and pollutant load can influence ecological processes. Modern usage of artificial fertilizers, containing nitrogen, phosphorus, and potassium, has affected ecosystems where rivers discharge into shallow seas. In the Yellow Sea and Bohai Sea of China, algae harmful to fish commonly form a bloom or 'red tide'. Similar blooms of harmful algae occur in the North Sea, where agricultural pesticides washed in by rivers cause problems for marine life.

Further reading: Gregory and Walling 1976

ECOLOGICAL NICHE

The idea of an ecological niche (or simply niche) is at once straightforward and complicated. In essence, an organism's ecological niche is its place in an **ecosystem** defined by its 'address' and 'profession'. Its address or home is the **habitat** in which it lives, sometimes called the habitat niche. Joseph Grinnell first suggested this idea in 1917. Its profession or occupation is its position or function in a food chain, and sometimes called the functional niche. Charles Elton developed this idea in 1927. A skylark's (*Alauda arvensis*) address is open moorland (and, recently, arable farmland); its profession is insect-cum-seed-eater. A merlin's (*Falco columbarius*) address is open country, especially moorland; its profession is a bird-eater, skylark and meadow pipit (*Anthus pratensis*) being its main prey. A grey squirrel's (*Sciurus carolinensis*) habitat niche is deciduous woodland; its profession is nut-eater (small herbivore). A grey wolf's (*Canis lupus*) habitat niche is cool temperate coniferous forest, and its profession is large-mammal-eater.

G. Evelyn Hutchinson (1957) defined a niche as a region (an *n*-dimensional hypervolume) in a multidimensional space of environmental factors that affect the welfare of a species, including temperature, moisture, and food size. This definition became popular because it is easier to establish the tolerance range of a species to ecological factors than it is to measure a species' 'profession'. Hutchinson distinguished between the fundamental niche and the realized niche. The fundamental (or virtual) niche circumscribes where an organism would live under optimal physical conditions and with no competitors or predators. The realized (or actual) niche is always smaller, and defines the 'real-world' niche occupied by an organism constrained by biotic and abiotic **limiting factors**.

A niche reflects how an individual, species, or **population** interacts with and exploits its **environment**. It involves **adaptation** to environmental conditions. The competitive exclusion principle precludes two species occupying identical niches. However, a group of species, or guild, may exploit the same class of environmental resources in a similar way (Root 1967). For instance, in oak woodland, one guild of birds forages for arthropods from the foliage of oak trees; another catches insects in the air; another eats seeds. The foliage-gleaning guild in a California oak woodland includes members of four families: the plain titmouse (*Parus inornatus*, Paridae), the blue-grey gnatcatcher (*Polioptila caerulea*, Sylviidae), the warbling vireo and Hutton's vireo (*Vireo gilvus* and *Vireo huttoni*, Vireonidae), and the orange-crowned warbler (*Vermivora celata*, Parulidae) (Root 1967).

Although only one species occupies each niche, different species may occupy the same or similar niches in different geographical regions. These species are ecological equivalents or vicars. A grassland ecosystem contains a niche for large herbivores living in herds. Bison and the pronghorn antelope occupy this niche in North America; antelopes, gazelles, zebra, and eland, in Africa; wild horses and asses, in Europe; the pampas deer and guanaco in South America; and kangaroos and wallabies in Australia. As this example shows, quite distinct species may become ecological equivalents through historical and geographical accidents. Many bird guilds have ecological equivalents on different continents. The nectar-eating (nectivore) guild has representatives in North America, South America, and Africa. In Chile and California, the representatives are the hummingbirds (Trochilidae) and the African representatives are the sunbirds (Nectariniidae). One remarkable convergent feature between hummingbirds and sunbirds is the iridescent plumage. Plant species of very different stock growing in different areas, when subjected to the same environmental pressures, have evolved the same life-form to fill the same ecological niche. The American cactus and the South African euphorbia, both living in arid regions, have adapted by evolving fleshy, succulent stems and by evolving spines instead of leaves to conserve precious moisture.

Further reading: Chase and Leibold 2003

ECOREGION

Canadian forester Orie Loucks first proposed the idea of an ecoregion (sometimes called a bioregion) in 1962. An ecoregion is a 'recurring pattern of **ecosystems** associated with characteristic combinations of **soil** and landform that characterise that region' (Brunckhorst 2000). The World Wildlife Fund (WWF) gives a longer definition: a large area of land or water with a geographically distinct assemblage of natural communities sharing a large majority of their species, ecological dynamics, environmental conditions, and interacting ecologically in ways that are critical for their long-term persistence. Its Conservation Science Program has identified 825 terrestrial ecoregions across the globe, and a set of about 450 freshwater ecoregions is under development. The WWF has recently launched an analogous global framework of 229 coast and shelf marine ecoregions in collaboration with The Nature Conservancy.

Further reading: Bailey 1995, 1996, 1997, 2002

ECOSYSTEM

In 1935, Arthur Tansley, a British ecologist, defined an ecosystem (ecological system) as a self-sustaining community of organisms together with the physical **environment** that supports it. However, evidence suggests that A. Roy Clapham suggested the term to Tansley in 1930 when asked if he could think of a suitable word to denote the physical and biological components of an environment considered in relation to each other as a unit (Willis 1994, 1997). Raymond Lindeman's (1942) groundbreaking paper in *Ecology* introduced the trophic–dynamic concept of ecosystems, which drew attention to the transfer of **energy** from one part of an ecosystem to another. It also categorized organisms into rather discrete trophic levels – producers, primary consumers (herbivores), secondary consumers (carnivores), and tertiary consumers (top carnivores) – each successively dependent on the preceding level as a source of energy. Lindeman (1942, 400) defined an ecosystem as 'the system composed of physical–chemical–biological processes active within a space–time unit of any magnitude'.

Ecosystems range in size from the smallest units that can sustain life (consisting of several species and a fluid medium) to the global ecosystem or ecosphere. Exchanges of energy and materials between living organisms and their supporting environment characterize them. The living organisms form an interacting set of microorganisms, animals, and plants organized into food chains, food webs, and trophic levels. The abiotic part consists of inorganic materials, organic by-products of biotic activity, and physical environmental factors (winds, tides, heat, light, and so on). Ecosystems are real units – the ecological community (the set of interacting species living in the same place) and its non-living environment functions as a unified, if complex and dynamic, whole. Central to the ecosystem concept is the notion that living organisms continually engage in a set of relationships with every other element constituting their environment. Extended to the global ecosystem, this notion is expressed in the aphorism that 'everything is connected to everything else'.

The wide acceptance of the ecosystem concept led to the rise of modern ecology and a framework for major international programmes, such as the International Biological Programme (IBP) and its successors.

Further reading: Dickinson and Murphy 2007

ECOTONE

An ecotone is a transition zone between two neighbouring **ecosystems**. It is normally evident in the changing distribution of vegetation, but it involves other organisms from the two ecosystems. An ecotone can be a gradual blending of two communities across many kilometres, as in the parkland ecotone between forest and savannah grassland in the subtropics. In these circumstances, the ecotone often contains animals and plants from both adjacent ecosystems, sometimes with species belonging uniquely to the ecotone ecosystem. In other cases, the change from one ecosystem to another can occur over a few hundred metres or less, as in some altitudinal ecotones, especially tree-lines, in mountain environments. Tree-lines are usually sharp ecotones in which a noticeable change of dominant life-form occurs. They sit at the circumpolar boundary between taiga and tundra vegetation in the Northern Hemisphere, and in the boundary between subalpine vegetation and low-growing alpine vegetation on mountains.

Changes in environmental factors, and particularly climate, **soils**, and geological substrate influence the position of many ecotones. For instance, location of the prairie–forest ecotone in the USA corresponds to a change from year-round runoff in the forested regions to spring runoff in the prairie. Fire, grazing intensity, the degree of competition between the two communities, and other factors may also play a role.

Ecotones are important for some animals, as they can exploit more than one set of **habitats** within a short distance. This can produce an edge effect along the boundary line, with the area displaying a greater than usual diversity of species.

ENDOGENIC (INTERNAL) FORCES

These are tectonic and volcanic forces generated deep inside the Earth by its vast heat engine. They affect the composition and dynamics of the Earth's surface **systems** – communities, **soils**, rivers and water bodies, landforms, and the atmosphere. Over very long periods, the immense endogenic forces often reveal at the surface such features as uplifted rocks or other forms of crustal deformation, folded mountains, and volcanoes.

Interestingly, all geomorphic systems in effect result from a basic antagonism between endogenic processes driven by geological forces and **exogenic** processes driven by climatic forces (Scheidegger 1979). In short, tectonic and volcanic processes create land and climatically influenced weathering and erosion destroy it. The events between the creation and the final destruction are what captivate geomorphologists.

ENERGY/ENERGY FLOW

Energy and energy flow are crucial to the understanding of many physical geographical processes. The idea of energy budgets has proved popular with physical geographers. An energy budget is an accounting of the energy flow through a system. Ecologists first developed energy budgets for ecosystems. Raymond Lindeman's (1942) trophic–dynamic model grouped organisms into feeding levels (what are now called producers, herbivores, carnivores, and top carnivores), each of which has an energy content or standing crop. Energy continuously enters and leaves a trophic level, so imparting a dynamic aspect to the ecosystem. A portion of the energy received by a trophic level dissipates as heat. From these theoretical beginnings evolved the subject of ecological energetics, pioneered by Eugene P. Odum and his brother Howard T. Odum.

Climatologists and meteorologists compute heat budgets of climate **systems**, considering the relationship between incoming solar (short-wave) radiation and outgoing terrestrial (long-wave) radiation. They do so at a range of scales, from global down to local and smaller. Globally, on the average, the Earth's surface is a heat source and the atmosphere is a heat sink, with the energy lost by the atmosphere matching the energy gained by the surface, so that the energy balance for the Earth–atmosphere system as a whole is, on the average, in balance with no overall gain or loss. However, there is a surplus of energy between about 40°N and 40°S, and a deficit poleward of those latitudes, and this inequality powers the **general circulation of the atmosphere**. Three processes transport the tropical energy surplus towards the temperate and polar regions of energy deficit – sensible heat flux, latent heat flux, and surface heat flux. The sensible heat flux is the transfer of energy from the Earth's surface to the atmosphere, which occurs through conduction and **convection**. Once in the atmosphere, sensible heat moves polewards by **advection**. The latent heat flux involves the advection and convection of water vapour, which releases its latent heat upon condensation. The tropical energy surplus also heats the tropical oceans, the energy in which warm ocean currents carry polewards. At a local scale, climatologists study the energetic basis of such phenomena as urban heat islands (e.g. Oke 1982) and topoclimates (e.g. Błazejczyk and Grzybowski 1993).

ENVIRONMENT

The idea of 'the environment' is hugely powerful and yet difficult to crystallize. In a narrow sense, the environment of a system is what lies outside its ambits: the Earth's environment is the Solar System and the rest of the Cosmos; the environment of a plant and animal community is the substrate it lives on and the air it lives in. In a broad sense, the environment is an integration of the diverse biological and physical elements that affect the life of organisms. In fact, there are a host of environments occurring at different geographical scales, commonly grouped according to their attributes – the aquatic environment, the marine environment, the terrestrial environment, and so forth. The term natural environment encompasses all living and non-living things occurring naturally on the Earth as a whole or part of it, as in the natural environment of Australia. It contrasts with the terms built environment and human environment, which includes such human constructed and dominated areas as towns and cities. Wilderness refers to areas without any human intervention, or nearly so, such as Antarctica.

Further reading: Head 2007

ENVIRONMENTAL CHANGE

Environmental change is a hugely potent idea. It is the subject of extensive cross-disciplinary research and a library of books and papers. Surprisingly perhaps, the fact of environmental change is not easy to establish. The problem lies in the natural variability of environmental factors. Take the example of air temperature. This varies on a daily, monthly, seasonal, yearly, and longer basis, and a yardstick, such as average temperature values for days, months, seasons, years, or longer periods, is required to measure sustained temperature changes. For instance, a comparison of annual averages may reveal that the mean temperature for one year is greater than that in the previous year – a change has occurred. There is a danger here of treating average values as some sort of norm. For example, average values for all the main atmospheric variables over 30-year periods define the misleadingly named 'climatic normals'. However, climatic normals themselves change: climate is always changing and one generation's climatic norms are another generation's climatic extremes. The tenor of this argument applies to all environmental factors – change is the norm, constancy the exception.

Environmental variables display three basic types of change – discontinuities, trends, and fluctuations (Figure 9). A discontinuity is an abrupt and permanent change in the average value. A trend is smooth increase or decrease, not necessarily a linear one, on the average. A fluctuation is regular or irregular change characterized by at least two maxima (or minima) and one minimum (or maximum). There are several kinds of fluctuation. An oscillation is smooth and gradual progress between maxima and minima. A periodicity is the recurrence of maxima and minima after a roughly constant time interval. Less than regular periodicity is quasi-periodicity. An episodicity is a sustained minimum (or 'norm') interrupted by an abrupt but tempo-rary switch to a maximum. It marks an event, such as a large flood or a landslide. Repeated events may occur randomly or quasi-periodically.

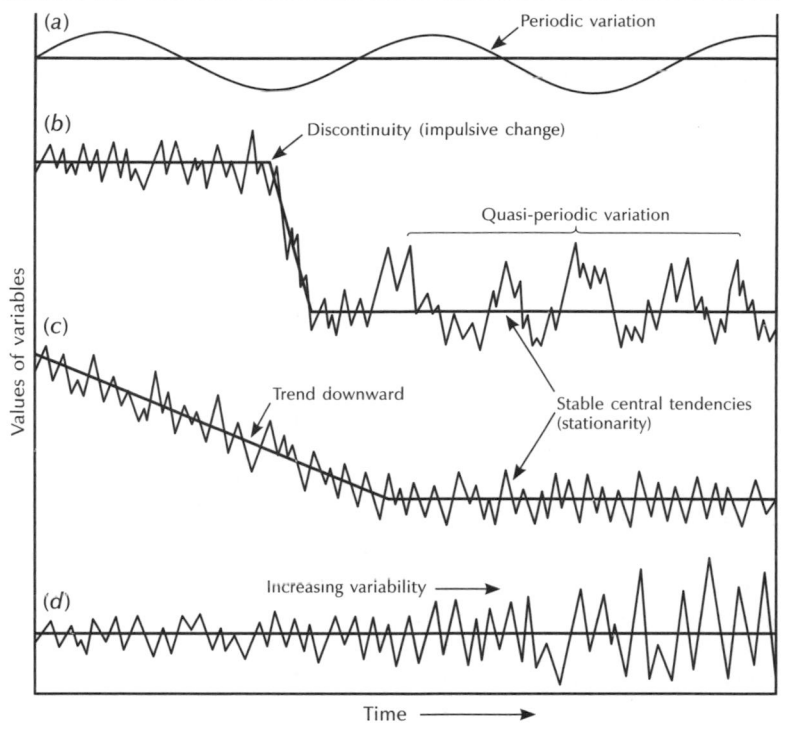

Figure 9 The chief styles of environmental change. The idealized time series apply to all variables that are continuous in time. Temperature and pressure are examples.

Source: Adapted from Hare (1996)

Another subtle possibility of change is that the average value is sustained for a time, while the variability about the average increases or decreases. Subtler still is chaotic change. Chaos describes a cryptic pattern of change so irregular that it is easily mistaken for randomness. All this sounds simple enough, but detecting discontinuities, trends, oscillations, and chaos in data sets is challenging.

Trends, or directional changes, are an important aspect of environmental change. There are two alternatives – the **environment** may change in a definite direction, or else it may stay the same. In the first case there is a changing or transient state; in the second case there is a steady state. Oscillations complicate this simple picture. Steady and changing environmental variables may follow cycles. Indeed, it is probably safe to say that all environmental variables display some degree of cyclical variation (see **cyclicity**).

An important aspect of environmental change is the rate at which it occurs. How fast does mean annual temperature increase? How rapidly does a glacier retreat? How quickly did trees colonize newly exposed land after deglaciation? Questions such as these are relatively easy to answer for the instrumental period. For historical and geological times, answers are perforce less sure. Indeed, before the coming of geochronometric dating methods, estimates of geological rates were at best a first approximation. Luckily, the absolute timescale kept by geochronometers help to gauge many geological rates with a high degree of confidence (see **geochronology**). Rates of environmental change range from slow to fast. Slow rates are usually referred to as gradual environmental change. Fast rates are sometimes described as catastrophic environmental change, though this term carries connotations of suddenness and violence and is shunned by many environmental, geographical, and geological scientists. Convulsive change is a euphemistic alternative. Nonetheless, catastrophes are common in the natural world, catastrophic change does occur, and there is no good reason for dropping a perfectly good word. All rates between the slow and fast extremes occur in the environment.

Further reading: Oldfield 2005; Huggett 1997a; Mannion 1997, 1999

EQUIFINALITY

Equifinality is the idea that different means can create the same end; in other words, two or more causes may produce the same effect. In more technical parlance, end states in open **systems** are reachable in more than one way. The term equifinality is due to Ludwig von Bertalanffy, the founder of General Systems Theory, from where Richard Chorley (1962) introduced it to physical geography. In closed systems, a direct cause-and-effect relationship exists between the initial state and the end state: if you push a switch, a light goes on. Open systems, which include all systems of interest to physical geographers, function differently, with similar end-states achievable from dissimilar initial conditions (starting points) and via different paths. A good example is the meandering pattern found in rivers, supraglacial streams (streams flowing upon glaciers), some ocean currents, and jet streams in the atmosphere. The starting states in all these cases differ, as to some extent do the processes involved, but the pattern produced (the end state) is the same.

Geomorphologists use the term equifinality (or convergence in the European literature) to indicate that different sets of processes can produce similar landforms. An example is tors, which may have formed in different ways, including tropical weathering and periglacial erosion. Another is patterned ground formation, the formation of which is complex, all the more so because similar kinds of patterned ground appear to be created by different processes (as in periglacial environments and in deserts), and the same processes can produce different kinds of patterned ground. Equifinality in geomorphology is not without problems, its recognition sometimes resulting from imprecision, rather than being truly present. Indeed, it perhaps better thought of as a situation where the same landforms result from the same set of processes acting upon different starting conditions (Haines-Young and Petch 1983).

In hydrology and ecology, some modelling studies suggest that two or more different models can satisfactorily mimic the behaviour of observed natural processes. The idea of equifinality is also useful in archaeology, where different historical processes may sometimes lead to a similar outcome, as in the development of agriculture, which seems to have occurred independently in different regions for different reasons and through different historical paths.

EQUILIBRIUM

Equilibrium is 'a condition in which some kind of balance is maintained' (Chorley and Kennedy 1971, 348). It is a complex concept that plays a starring role in all branches of physical geography, chiefly perhaps because it is highly significant when considering change though time. Its complexity lies in the multiplicity of equilibrial patterns and the fact that not all components of a system need be in balance at the same time for some form of equilibrium to obtain.

Figure 10 shows eight conditions of equilibrium (a–h). Static equilibrium is the condition where an object has forces acting upon it but it does not move because the forces balance, an example being a boulder resting on a slope. Thermodynamic equilibrium is the tendency towards maximum entropy, as demanded by the second law of thermodynamics. Stable equilibrium is the tendency of a system to revert to its original state after experiencing a small perturbation, as when a mammal's body temperature adjusts to a change in environmental temperature. Unstable equilibrium occurs when a small perturbation forces a system away from its old equilibrium state towards a new one. This might occur in **systems** with multiple or alternative stable states, as in some natural communities (Sutherland 1974; Temperton *et al.* 2004). Metastable equilibrium involves a stable

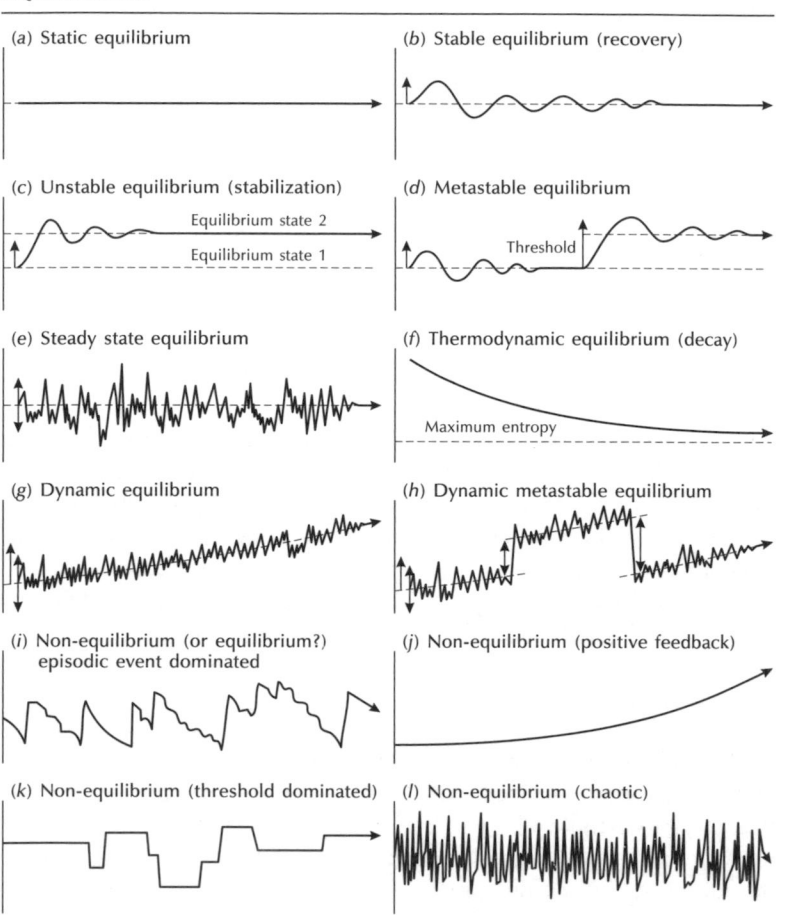

Figure 10 Types of equilibrium in geomorphology.

Source: Adapted from Chorley and Kennedy (1971, 202) and Renwick (1992)

equilibrium acted upon by some form of incremental change (a trigger mechanism) that drives the system over a **threshold** into a new equilibrium state. A stream, for instance, if forced away from a steady state, will adjust to the change, although the nature of the adjustment may vary in different parts of the stream and at different times. Douglas Creek in western Colorado, USA, was subject to overgrazing during the 'cowboy era' and, since about 1882, it has cut into its channel bed (Womack and Schumm 1977). The manner of cutting has been complex, with discontinuous episodes of downcutting interrupted by phases of deposition, and with the erosion–deposition sequence varying from one cross-section to another. Steady-state equilibrium occurs when numerous small-scale fluctuations occur about a mean stable state. The notion of steady state is perhaps the least controversial of systems concepts in physical geography. Any open system may eventually attain time-independent equilibrium state – a steady state – in which the system and its parts are unchanging, with maximum entropy and minimum free **energy**. In such a steady state, a system stays constant as a whole and in its parts, but material or energy continually passes through it. As a rule, steady states are irreversible. Before arriving at a steady state, the system will pass through a transient state (a sort of start-up or warm-up period). For instance, the amount of water in a lake could remain steady because gains of water (incoming river water and precipitation) balance losses through river outflow, groundwater seepage, and evaporation. If the lake started empty, then its filling up would be a transient state. Dynamic equilibrium is a disputatious term and is discussed in the next paragraph. Dynamic metastable equilibrium is a combination of dynamic and metastable equilibria, in which large jumps across thresholds break in upon small-scale fluctuations about a trending mean. Some of these so-called types of equilibrium are really forms of disequilibrium as a progressive change of the mean state occurs, as in dynamic metastable equilibrium (Renwick 1992). Figure 10 also shows four types of non-equilibrium (i–l), which range from a system lurching from one state to another in response to episodic threshold events to a fully chaotic sequence of state changes.

The term dynamic equilibrium perhaps demands a fuller discussion. Chemists first used the expression dynamic equilibrium to mean equilibrium between a solid and a solute maintained by solutional loss from the solid and precipitation from the solution running at equal rates. The word equilibrium captured that balance and the word dynamic captured the idea that, despite the equilibrium state, changes take place. In other words, the situation is a dynamic, and not a static, equilibrium. Grove Karl Gilbert (1877) possibly first applied the term in this sense in a physical geographical context. He suggested that all streams work towards a graded condition, and attain a state of dynamic equilibrium when the net effect of the flowing water is neither the erosion of the bed nor the deposition of sediment, in which situation the landscape then reflects a balance between force and resistance. Applied to any landform, dynamic equilibrium would represent a state of balance in a changing situation. Thus, a spit may appear to be unchanging, although deposition feeds it from its landward end, and erosion consumes it at its seaward end. John T. Hack (1960) developed Gilbert's ideas, arguing that a landscape should attain a steady state, a condition in which land-surface form does not change despite material being added by tectonic uplift and removed by a constant set of geomorphic processes. He contended that, in an erosional landscape, dynamic equilibrium prevails where all slopes, both hillslopes and river slopes, are adjusted to each other (cf. Gilbert 1877, 123–24; Hack 1960, 81), and 'the forms and processes are in a steady state of balance and may be considered as time independent' (Hack 1960, 85). In practice, this notion of dynamic equilibrium was open to question (e.g. Ollier 1968) and difficult to apply to landscapes. In consequence, geomorphologists advanced other forms of equilibrium, of which dynamic metastable equilibrium has proved to be salutary. Other physical geographers have used the term dynamic equilibrium to mean 'balanced fluctuations about a constantly changing system condition which has a trajectory of unrepeated states through time' (Chorley and Kennedy 1971, 203), which is similar to Alfred J. Lotka's (1924) idea of moving equilibrium (cf. Ollier 1968, 1981, 302–4). Currently then, dynamic equilibrium in physical geography is synonymous with a 'steady state' or with a misleading state, where the system appears to be in equilibrium but in reality is changing extremely sluggishly. Thus, the term has been a replacement for concepts such as grade and climax. Problems with the concept relate to the application of a microscale phenomenon in physics to macroscale physical geographical systems, and to the difficulty of separating any observed fluctuations from a theoretical underlying trend (Thorn and Welford 1994). On balance, it is perhaps better for physical geographers to abandon the notion of dynamic equilibrium, and indeed some of the other brands of equilibrium, and instead adopt the terminology of nonlinear dynamics (see **complexity**).

Further reading: Bracken and Wainwright 2006

ERGODICITY (SPACE–TIME OR LOCATION–TIME SUBSTITUTION)

Ergodicity is an idea developed in physics and adopted by geomorphology. Most geomorphological applications shun the rigorous assumptions made by physicists and simply use space–time (or better, location–time) substitution, which involves identifying similar landforms of differing age at different locations, and then arranging them chronologically to create a time sequence or topographic **chronosequence**. Such research has proved salutary in understanding landform development. Two broad types of location–time substitution are used. The first looks at **equilibrium** ('characteristic') landforms and the second looks at non-equilibrium ('relaxation') landforms.

In the first category of location–time substitution, the assumption is that the geomorphic processes and forms under consideration are in equilibrium with landforms and environmental factors. For instance, modern rivers on the Great Plains of the USA display relationships between their width–depth ratio, sinuosity, and suspended load, which aid the understanding of channel change through time (Schumm 1963). Allometric models are a special case of this kind of location–time substitution (see Church and Mark 1980). Studies in the second category of location–time substitution, which look at developing or 'relaxation' landforms, bear little relationship to the ergodicity of physics. The argument is that similar landforms of different ages occur in different places. A developmental sequence emerges by arranging the landforms in chronological order. The reliability of such location–time substitution depends upon the accuracy of the landform chronology. Least reliable are studies that simply assume a time sequence. Charles Darwin, investigating coral-reef formation, thought that barrier reefs, fringing reefs, and atolls occurring at different places represented different evolutionary stages of island development applicable to any subsiding volcanic peak in tropical waters. William Morris Davis applied this evolutionary schema to landforms in different places and derived what he deemed was a time sequence of landform development – the **geographical cycle** – running from youth, through maturity, to senility. This seductively simple approach is open to misuse. The temptation is to fit the landforms into some preconceived view of landscape change, even though other sequences might be constructed. More useful are situations where, although an absolute chronology is unavailable, field observations enable geomorphologists to place the landforms in the correct order. This occasionally happens when, for instance, adjacent hillslopes become progressively removed from the action of fluvial or marine processes at their bases. This has happened along a segment of the South Wales coast, in the British Isles, where a sand spit growing from west to east has affected the Old Red Sandstone cliffs between Gilman Point and the Taff estuary (see p. 31). Relative-age chronosequences depend upon some temporal index that, though not fixing an absolute age of landforms, enables the establishment of an interval scale. For example, the basin hypsometric integral and stream order both measure the degree of fluvial landscape development and are surrogates of time (e.g. Schumm 1956). The most informative examples of location–time substitution arise where absolute landform chronologies exist (see **geochronology**). Historical evidence of slope profiles along Port Hudson bluff, on the Mississippi River in Louisiana, southern USA, revealed a dated chronosequence (Brunsden and Kesel 1973). The Mississippi River was undercutting the entire bluff segment in 1722. Since then, the channel has shifted about 3 km downstream with a concomitant stopping of undercutting. The changing conditions at the slope bases have reduced the mean slope angle from 40° to 22°.

Location–time substitution does have pitfalls. First, not all spatial differences are temporal differences because factors other than time exert an influence on landforms. Second, landforms of the same age might differ through historical accidents. Third, **equifinality**, the idea that different sets of processes may produce the same landform, may cloud interpretation. Fourth, process rates and their controls may have changed in the past, with human impacts presenting particular problems. Fifth, equilibrium conditions are unlikely to have endured for the timescales over which the locational data substitute for time, especially in areas subject to Pleistocene glaciations. Sixth, some ancient landforms are relics of past environmental conditions and are in disequilibrium with present conditions. Despite these latent problems, geomorphologists substitute location for time to infer the nature of landform change, and the loose application of ergodic reasoning is a productive line of geomorphological enquiry.

Further reading: Burt and Goudie 1994; Thornes and Brunsden 1977 (pp. 19–27); Paine 1985

ETCHPLANATION

Traditional models of landscape **evolution** assumed that mechanical erosion is far more important a factor than chemical erosion. Geomorphologists realized that chemical weathering reduces the mass of weathered material, but they argued that only on rocks especially vulnerable to solution (such as limestones) would chemical processes have an overriding influence on landscape evolution. However, it now seems that forms of chemical weathering are important in the evolution of many landscapes. In tropical and subtropical environments, chemical weathering produces a thick regolith that erosion then strips. This process is etchplanation and it fashions an etched plain or etchplain. The etchplain is largely a production of chemical weathering. In places where the regolith is deeper, weakly acid water lowers the weathering front, in the same way that an acid-soaked sponge would etch a metal surface. Some researchers contend that surface erosion lowers the land surface at the same rate that chemical etching lowers the weathering front (Figure 11). This is the theory of double planation. It envisages land surfaces of low relief maintained during prolonged, slow uplift by the continuous lowering of double plana-tion surfaces – the wash surface and the basal weathering surface (Büdel 1957; Thomas 1965). A rival view is that a period of deep chemical weathering precedes a phase of regolith stripping (e.g. Linton 1955; Ollier 1959, 1960).

Whatever the details of the etching process, it is very effective in creating landforms, even in regions lying beyond the present tropics. The Scottish Highlands experienced a major uplift in the Early Tertiary. After 50 million years, the terrain evolved by dynamic etching with deep weathering of varied geology under a warm to temperate humid climate (Hall 1991). This etching led to a progres-sive differentiation of relief features, with the evolution of basins, valleys, scarps, and inselbergs. In like manner, etchplanation may have played a basic role in the Tertiary **evolutionary geomorphology** of the southern England Chalklands, a topic that has always generated much heat. There is a growing recognition that the fundamental erosional surface is a summit surface formed by etchplanation during the Palaeogene period, and is not a peneplain formed during the Miocene and Pliocene periods (Jones 1999).

Further reading: Twidale 2002

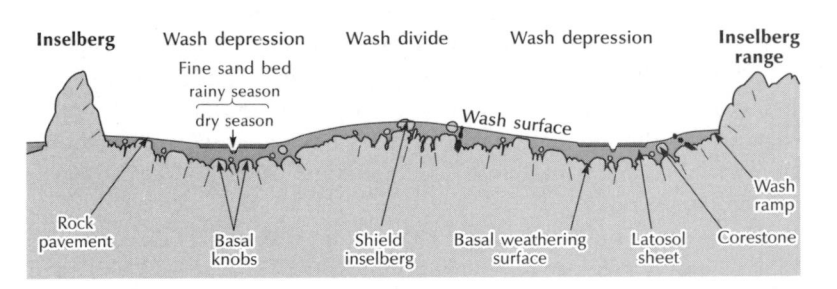

Figure 11 Double planation surfaces: the wash surface and the basal-weathering surface.

Source: Adapted from Büdel (1982, 126)

EUSTASY

The term eustasy refers to a uniform worldwide **sea-level change**. Water additions or extractions from the oceans (glacio-eustatic change) and changes in ocean-basin volume (tectono-eustatic change) lead to eustatic change (Table 4). In addition, steric change, associated with the temperature or density changes in seawater, also affects global sea level. Much of the predicted sea-level rise during the current century will result from the thermal expansion of seawater.

Climate change is the chief driver of glacio-eustatic change. Globally, inputs from precipitation and runoff normally balance losses from evaporation. (Gains from juvenile water probably balance losses in buried connate water.) However, when the climate system switches to an icehouse state, ice sheets and glaciers lock up a substantial portion of the world's water supply. Sea level drops during glacial stages, and rises during interglacial stages. Additions and subtractions of water from the oceans, other than that converted to ice, may cause small changes in ocean volume.

Geological processes drive tectono-eustatic change. Even when the water cycle is in a steady state, so that additions from precipitation balance losses through evaporation, sea level may change owing to volumetric changes in the ocean basins. An increasing volume of ocean basin would lead to a fall of sea level and a decreasing volume to a rise of sea level. Sedimentation, the growth of mid-ocean ridges, and Earth expansion (if it should have occurred) decrease ocean basin volume; a reduced rate or cessation of mid-ocean ridge production increases ocean basin volume.

Geoidal eustasy results from processes that alter the Earth's equipotential surface, or geoid. The relief of the ocean geoid (geodetic sea level) is considerable: there is a 180-m sea-level difference between the rise at New Guinea and the depression centred on the Maldives, which places lie a mere 50°–60° of longitude from one another. There is also a geoid beneath the continents. The configuration of the geoid depends on the interaction of the Earth's gravitational and rotational potentials. Changes in geoid relief are often rapid and lead to swift changes of sea level.

On a short timescale, local changes in weather, hydrology, and oceanography produce relatively minor fluctuations of sea level, up to 5 m for major ocean currents, but less than half that for meteorological and hydrological changes.

Further reading: Dott 1992

Table 4 Causes of eustatic change

Seat of change	Type of change	Approximate magnitude of change (m)	Causative processes
Ocean basin volume	Tectono-eustatic	50–250	Orogeny Mid-ocean ridge growth Plate tectonics Sea-floor subsidence Other Earth movements
Ocean water volume	Glacio-eustatic Hydro-eustatic	100–200 Minor	Climatic change Change in liquid hydrospheric stores (water in sediments, lakes, clouds) Additions of juvenile water Loss of connate water
Ocean mass distribution and surface 'topography'	Geoidal eustatic	Up to 18 A few metres 1 per millisecond of rotation Up to 5 2 (during Holocene)	Tides Obliquity of the ecliptic Rotation rate Differential rotation Deformation of geoid relief
	Climo-eustatic	Up to 5 for major ocean currents	Short-term meteorological, hydrological, and oceanographic changes

Source: Partly adapted from Mörner (1987, 1994)

EVOLUTION

The idea of evolution underpins much of physical geography, especially in those branches of the discipline dealing with historical aspects of the **environment**. Regrettably, confusion surrounds the term because it has many meanings. Literally speaking, evolution means 'unfolding' or 'unrolling' and in biology it is understood in two chief ways (Mayr 1970). First, evolution means the unfolding, or growth and development, of an individual organism. This process of ontogeny involves **homeorhesis** (Waddington 1957). Second, in a grander sense, evolution means phylogenetic evolution – the derivation of all life-forms from a single common ancestor. There is a crucial difference between these two ideas. Development (homeorhesis) produces a new organism that is almost identical to its progenitors (or identical in the case of asexual reproduction). Phylogenetic evolution creates organisms that have never before existed, and that may be more complex than their progenitors. A process of complexification occurs in both cases. With development, complexification leads to a familiar, pre-existing organism; with phylogenetic evolution, complexification leads to a novel organism, often at some higher level of organization. These then, briefly, are the biological meanings of 'evolution'.

As well as biological evolution (the evolution of living things), there are inorganic evolution (cosmic and planetary evolution) and psychological evolution (the evolution of humans and their cultures) (Huxley 1953). A crucial question here is in what sense do non-biological **systems**, such as **soils** and landforms, evolve? Modern physical geography has inherited a developmental view of landforms, soils, and communities from late nineteenth and early twentieth century thinkers. The argument runs that non-biological systems (including animal and plant communities) follow, time after time, predetermined developmental (homeorhetic) pathways. In geomorphology, for instance, landscape development is, according to William Morris Davis, a process of reduction. It always has, and always will, progress through the same developmental sequence – youth, maturity, and old age – to end with an all-but level plain (see **geographical cycle**). In pedology, the traditional view (soil formation theory, see **pedogenesis**) is that soil forms or develops progressively under the influence of the environmental state factors. The developmental sequence continues until the soil is in **equilibrium** with prevailing environmental conditions, whence the soil is 'mature' (e.g. a podzol or chernozem) and will change no more. In ecology, vegetation develops through successive seral stages until it attains a 'mature' or climax state (such as steppe grassland and deciduous forest) that is in balance with environmental, and especially climatic, conditions (see **climax community**). Plainly, according to the developmental view, landforms, soils, and communities will always change in predictable ways, following the uplift or exposure of new land. The sequence of change envisaged is homeorhetic in character, and ends with a sort of permanent **homeostasis** (in the mature or climax stage).

Recent empirical work on **environmental change** undermines the tenets of the developmental view, for it shows that inconstancy, not constancy, of environmental conditions is the norm. Considering this fact, it is improbable that a developmental sequence of landforms, soils, or vegetation will ever run its full course under a constant environment. To complicate matters further, some researchers find it profitable to regard all environmental systems (including climate, communities, landforms, landforms, and soils) as dissipative structures (see **system**) replete with nonlinear relations and forced away from equilibrium states by driving variables. This nonlinearity in systems removed from equilibrium may generate chaotic regimes. After having entered a chaotic regime, internal dynamics and **thresholds** drive landforms, soil, and communities through a series of fundamentally unpredictable states. The nature of these states is strongly dependent on the initial conditions. This is the antithesis of the developmental view, in which the initial state is considered unimportant. Environmental inconstancy and nonlinear dynamics lead, then, to a far more dynamic picture of environmental change than early generations of scientists could scarcely have imagined. The systems of the ecosphere are generally plastic in nature and respond to changes in their environment and to thresholds within themselves. The result is that climates, communities, landscapes, and soils evolve, rather than develop (cf. Huggett 1995). Their genesis involves continual creation and destruction at all scales, and may progress or retrogress depending on the environmental circumstances. They do not seem inevitably to pursue a preordained developmental path. Rather, they constantly evolve, responding to continual changes in their internal, cosmic, and geological environments. This evolutionary view of environmental change is of the utmost significance. It means that at any instant, environmental systems are unique and changing, and greatly contingent upon historical events (owing to the relevance of initial conditions). This is the basis of **evolutionary geomorphology** and **evolutionary pedology**. An evolutionary view makes it very difficult to predict change. Landforms, soils, and communities (and probably climates) formed under the same environmental constraints are likely to be broadly similar, but they will invariably differ in detail. In short, an evolutionary view offers a new way of thinking about and studying environmental change.

Further reading: Futuyma 2005; Milner 1990; Ridley 2003

EVOLUTIONARY GEOMORPHOLOGY

The non-actualistic system of land-surface history known as evolutionary geomorphology (Ollier 1981, 1992) makes explicit directional change in landscape development. The argument runs that the land surface has changed in a definite direction through time, and has not suffered the 'endless' progression of erosion cycles first suggested by James Hutton and implicit in William Morris Davis's **geographical cycle**. An endless repetition of erosion cycles would simply maintain a steady state with Silurian landscapes looking very much like Cretaceous landscapes and modern landscapes. Evolutionary geomorphologists contend that the Earth's landscapes have evolved as a whole. In doing so, they have been through several geomorphological 'revolutions', which have led to distinct and essentially irreversible changes of process regimes, so that the nature of erosion cycles has changed with time. These revolutions probably occurred during the Archaean aeon, when the atmosphere was reducing rather than oxidizing, during the Devonian period, when a cover of terrestrial vegetation appeared, and during the Cretaceous period, when grassland appeared and spread (see **actualism**).

The breakup and coalescence of continents would also alter landscape **evolution**. The geomorphology of Pangaea was, in several respects, unlike present geomorphology (Ollier 1991, 212). Vast inland areas lay at great distances from the oceans, many rivers were longer by far than any present river, and terrestrial sedimentation was more widespread. When Pangaea broke up, rivers became shorter, new continental edges were rejuvenated and eroded, and continental margins warped tectonically. Once split from the supercontinent, each Pangaean fragment followed its own history; each experienced its own unique events. These included the creation of new plate edges and changes of latitude and climate. It also involved substantial changes in drainage systems (Beard 2003; Goudie 2005). The landscape evolution of each continental fragment must be viewed in this very long-term perspective. In this evolutionary context, the current fads and fashions of geomorphology – process studies, dynamic **equilibrium**, and cyclical theories – have limited application (Ollier 1991, 212).

A good example of evolutionary geomorphology is afforded by **tectonics** and landscape evolution in south-east Australia (Ollier and Pain 1994; Ollier 1995). Morphotectonic evolution in this area appears to represent a response to unique, non-cyclical events. Today, the Canobolas and Victoria Divides, which are intersected by the Great Divide and putative Tasman Divide to the east separate three major basins – the Great Artesian Basin, the Murray Basin, and the Gippsland–Otway basins (Figure 12). These divides are major watersheds. They evolved in several stages from an initial Triassic palaeoplain sloping down westwards from the Tasman Divide (Figure 13).

First, the palaeoplain was downwarped towards the present coast, forming an initial divide. Then the Great Escarpment formed and retreated westwards, facing the coast. Much of the Great Divide is at this stage. Retreat of slopes from the coast and from inland reduced the palaeoplain to isolated high plains, common on the Victoria Divide. Continued retreat of the escarpment consumed the high plains and produced a sharp ridge divide, as is seen along much of the Victoria Divide. The sequence from low-relief palaeoplain to knife-edge ridge is the reverse of peneplanation. With no further tectonic complications, the present **topography** would presumably end up as a new lower-level plain. However, the first palaeoplain is Triassic in age and the 'erosion cycle' is unlikely to end given continuing tectonic changes to interrupt the erosive processes. The morphotectonic history of the area is associated with unique events. These include the sagging of the Murray Basin, the opening of the Tasman Sea and creation of a new continental margin, the eruption of the huge Monaro volcano, and the faulting of huge blocks in Miocene times. The geomorphology is evolving, and there are no signs of erosional cycles or steady states.

Further reading: Phillips 2006b

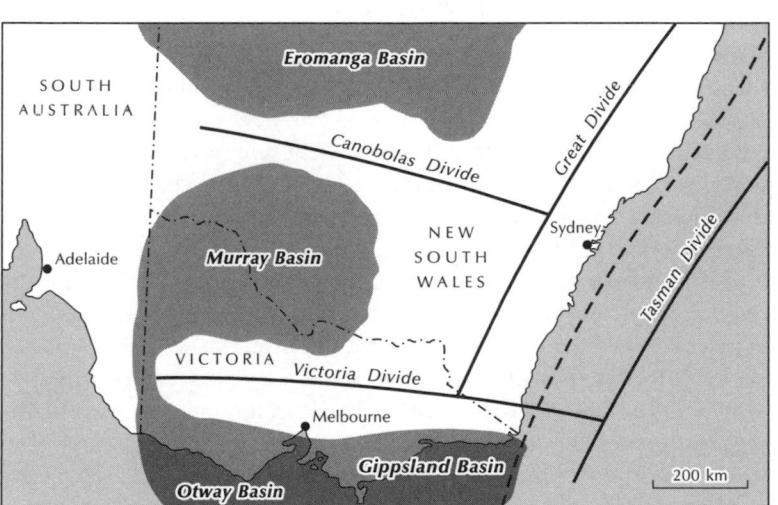

Figure 12 Major basins and divides of south–east Australia. The Eromanga Basin is part of the Great Artesian Basin.

Source: Adapted from Ollier (1995)

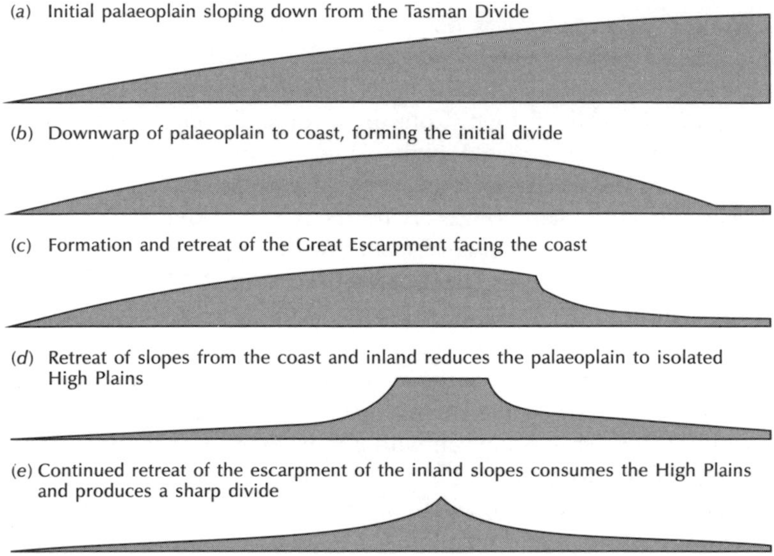

(a) Initial palaeoplain sloping down from the Tasman Divide

(b) Downwarp of palaeoplain to coast, forming the initial divide

(c) Formation and retreat of the Great Escarpment facing the coast

(d) Retreat of slopes from the coast and inland reduces the palaeoplain to isolated High Plains

(e) Continued retreat of the escarpment of the inland slopes consumes the High Plains and produces a sharp divide

Figure 13 Evolution of the south-east Australian drainage divides.

Source: Adapted from Ollier (1995)

EVOLUTIONARY PEDOLOGY

Soils and weathering features have evolved over geological time. Most of this **evolution** involved the appearance of new soil types and novel weathering features (Figure 14). The new soils evolved in new environments and new ecosystems. Ancient soils have persisted alongside newly evolved soils throughout Earth history. In consequence, soil diversity has increased in parallel with mounting environmental diversity, though progenitors of all modern soils orders (except for Mollisols, which are unknown since the Eocene epoch) existed during the Palaeozoic era (Figure 14). In theory, soils could go 'extinct', but in practice, only the Green Clay palaeosols from the early Precambrian no longer exist (Retallack 2001). Green Clay palaeosols formed in iron–rich parent materials such as basalt. They are, as their name suggests, green and clayey. They are also rich in alumina and poor in bases. Basalts currently weather to iron–rich red soils. The Green Clay palaeosols evolved under a weathering regime with very low atmospheric oxygen levels. Extinct soils appear to be a rarity. This could be because it is easier to recognize a palaeosol that has a modern counterpart, even when it has changed during burial, than it is a unique palaeosol.

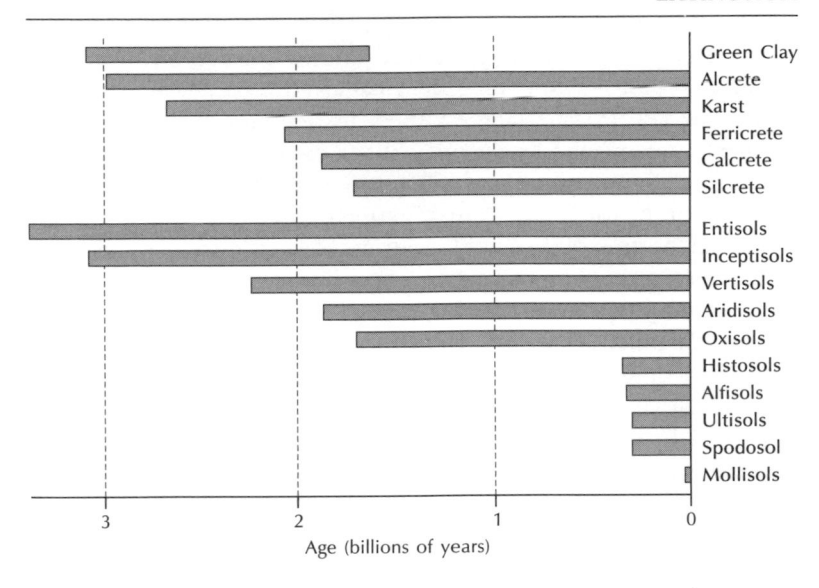

Figure 14 Geological time ranges of soil orders and weathering features.

Source: Adapted from Retallack (1986)

Further reading: Retallack 2001, 2003

EXOGENIC (EXTERNAL) FORCES

Geomorphic or exogenic agents are wind, water, waves, and ice, which act from outside or above the land surface. Gravity and solar **energy** are their prime divers. Although exogenic forces may be mundane or even minuscule compared to some **endogenic forces**, they are capable of wearing down, whether by gravity, water, wind, or ice, anything that endogenic forces can erect – there is no rock too resistant, no mountain too mighty, that exogenic agents cannot destroy given sufficient time.

EXTINCTION

Extinction is the eventual doom of all species (or genera, families, and so on). It may occur locally, globally, or on a massive scale. A local extinction or extirpation is the loss of a species from a particular place, but some of the gene pool survive elsewhere. The California condor (*Gymnogyps californianus*) is now extinct over much of its former range, but survives in southern California, Baja California, and the Grand

Canyon, where it was reintroduced after a captive breeding programme. A global extinction is the total loss of a particular gene pool.
When the last *Tyrannosaurus rex* died, its unique gene pool was lost
forever. Supraspecific groups may suffer extinctions. An example is
the global extinction of the sabre-toothed cats, one of the main
branches of the cat family. A mass extinction is a catastrophic loss of a
substantial portion of the world's species. Mass extinctions stand out in
the fossil record as times when the extinction rate runs far higher than
the background or normal extinction rate. However, some 99.99 per
cent of all extinctions are normal extinctions. Several factors may cause
mass extinctions, including asteroid and comet impacts, radiation from
supernovae, large solar flares, geomagnetic reversals, **continental
drift** (with associated **climate changes**), volcanism, **sea-level
change**, salinity change, anoxia and hypoxia, methane hydrate release,
and diseases.

Normal extinctions depend on many interrelated factors that fall
into three groups – biotic, evolutionary, and abiotic. The action of
most biotic factors of extinction depends upon **population** size (or
density): the larger the population, the more effective is the factor.
Density-dependent factors are chiefly biotic in origin. They include
factors related to biotic properties of individuals and populations (body
size, niche size, range size, population size, generation time, and
dispersal ability) and factors related to interactions with other species
(competition, disease, parasitism, and predation). As a rule, large
animals are more likely to become extinct than small animals. Smaller
animals can probably better adapt to small-scale **habitat** changes that
follow **environment** changes. Large animals cannot so easily find
suitable habitat or food resources and so find it more difficult to
survive. Specialist species with narrow niches are more vulnerable to
extinction than are generalists with wide niches. Small populations are
more prone to extinction though chance events, such as droughts,
than are large populations. In other words, there is safety in numbers.
Species with rapid generation times stand more chance of dodging
extinction. Good dispersers stand more chance of escaping extinction
than poor dispersers do, as do species with better opportunities for
dispersal. In addition, a species with a large gene pool may be better
able to adapt to **environmental change**s than species with a small
gene pool. Geography can be important – widespread species are less
likely to go extinct than species with restricted ranges. This is because
restricted range species are more vulnerable to chance events, such as a
severe winter or drought. In a widespread species, severe events may
cause local extinctions but are not likely to cause a global extinction.

Widespread species also appear to be less at risk than restricted species to mass extinctions. Competition can be a potent force of extinction. Species have to evolve to outwit their competitors, and a species that cannot evolve swiftly enough is in peril of becoming extinct. Virulent pathogens, such as viruses, may evolve or arrive from elsewhere to destroy species. The fungus *Phiostoma ulmi*, which is carried mainly by the Dutch elm beetle (*Scolytus multistriatus*), causes Dutch elm disease. Predators at the top of food chains are more susceptible to a loss of resources than are herbivores lower down. A chief factor in the decline of tigers is not **habitat loss** or poaching, but a depletion of the ungulate prey base throughout much of the tigers' range (Karanth and Stith 1999). Island mammal, bird, and reptile populations are especially vulnerable to all sorts of competitive and predatory introduced species. Since 1600 (and up to the late 1980s), 113 species of birds have become extinct. Of this total, 21 were on mainland areas and 92 on islands (Reid and Miller 1989). In many cases, numerous species of sea birds survive only on outlying islets where introduced species have failed to reach. The story for mammals and reptiles is similar.

Several chance evolutionary changes may lead to some species being more prone to extinction than others are. Evolutionary blind alleys arise when a loss of genetic diversity during **evolution** fixes species into modes of evolutionary development that become lethal. A species may evolve on an island and not possess the dispersal mechanisms to escape if the island should be destroyed or should experience climatic change. Some species may become overspecialized through **adaptation** and fall into evolutionary traps. Faced with **environmental change**, overspecialized species may be unable to adapt to the new conditions, their overspecialization serving as a sort of evolutionary straitjacket that keeps them 'trapped'. An interesting upshot of this idea is that species alive today must be descendants of non-specialized species. Behavioural, physiological, and morphological complexity, as varieties of specialization, also appear to render a species more prone to extinction. Simple species – marine bivalves for example – survive for about 10 million years, whereas complex mammals survive for 3 million years or less.

Abiotic factors of extinction act uniformly on populations of any size; that is, they are usually density-independent factors. Density-independent factors tend to be physical in origin – **climate change**, sea-level change, flooding, asteroid and comet impacts, and other catastrophic events. These factors often produce fluctuations in population size that can end in extinction. Abiotic factors are the most likely culprits for mass extinctions. However, several researchers stress the potential role of diseases as drivers of large-scale extinctions. Lethal pathogens carried by the dogs, rats, and other animals associated with migrating humans may have contributed to the Pleistocene mass extinctions (MacPhee and Marx 1997; Lyons *et al.* 2004).

Further reading: Lawton and May 1995; Benton 2003; Erwin 2006; Huggett 2006; Ward 2007

FEEDBACK

The terms negative and positive feedback, the broad applicability of which was recognized by Norbert Wiener (1948) in his book on cybernetics, are commonplace in the physical geographical literature, attesting to their allure and efficacy as concepts about system dynamics. Negative feedback occurs when a change in a system sets in motion a sequence of changes that eventually neutralize the effects of the original change, so stabilizing the system. An example occurs in a **drainage basin** system, where increased channel erosion leads to a steepening of valley-side slopes, which accelerates slope erosion, which increases stream bed-load, which reduces channel erosion (Figure 15). The reduced channel erosion then stimulates a sequence of events that stabilizes the system and counteracts the effects of the original change. In **population** biology, the growth of a prey population tends to increase the size a predator population, and the increased predator population then reduces the prey population, and so on. Positive feedback occurs when an original change in a system grows and the system becomes unstable. An example is an eroding hillslope where the slope erosion causes a reduction in infiltration capacity of water, which increases the amount of surface runoff, which promotes even more slope erosion (Figure 15).

In many environmental **systems**, a host of system feedback relationships, some negative and some positive, ring the system changes. A system as a whole may display a steady state, implying a preponderance of negative feedback, or it may grow or decline, implying the supremacy of positive feedbacks. There is a danger of mentally associating negative feedback relationships with a stable system, and positive feedback with a system moving away from stability. Without doubt, it is wrong to equate all positive feedback with the idea of a vicious circle. Admittedly, the example of positive feedback on an eroding hillslope did constitute a vicious circle, whereby the destabilized system continues changing. However, in biological and ecological

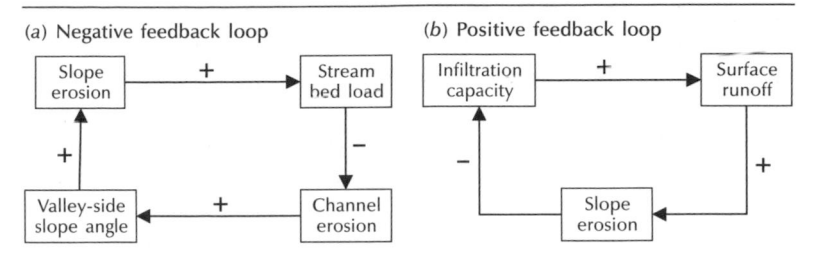

Figure 15 Feedback relationships in geomorphic systems. (a) Negative feedback in a valley-side slope–stream system. (b) Positive feedback in an eroding hillslope system. Details of the relationships are given in the text.

Source: After Huggett (2007a)

systems, positive feedback is fundamental to all growth processes where it creates a virtuous circle.

Negative feedback and positive feedback are valuable ideas. However, to an extent, the new views on systems dynamics and stability presented in the notion of **complexity** have supplanted them.

FUNCTIONAL–FACTORIAL APPROACH

The functional–factorial approach to **soil** and **ecosystem** genesis, which had a far-reaching impact in many environmental sciences, was a ruling theory in pedology, and to some extent ecology, until recently (see Johnson and Hole 1994). Its premise is that environmental factors determine the characteristics of soil and vegetation, and that mathematical functions can relate soil and vegetation characteristics to the environmental factors. Hans Jenny (1941) first influentially expressed this idea as the 'clorpt' equation, which provides a conceptual and analytical framework for studying interactions between components of the ecosphere, and especially soils and ecosystems. The 'clorpt' equation states that any soil property, s, is a function of soil-forming factors:

$$s = f(cl, o, r, p, t, \ldots)$$

where cl is environmental climate; o is organisms (the fauna and flora originally in the system and that entering later); r is **topography**, also including hydrological features such as the water table; p is parent material, defined as the initial state of soil when **pedogenesis** starts;

t is the age of the soil, or absolute period of soil formation; and the dots are additional factors such as fire. Jenny (1980, 203) opined that the 'clorpt' equation is a synthesis of information about land ecosystems. He suggested that, in favourable landscapes, researchers could assess the effect of one state factor on a single ecosystem property, all other factors held constant. Given five state factors, Jenny proposed five broad groups of functions or sequences: climofunctions or climosequences, biofunctions or biosequences, topofunctions or toposequences, lithofunctions or lithosequences, and chronofunctions or **chronosequences**. He also included dotfunctions and dotsequences to allow for the effects of other factors such as fire.

The 'clorpt' equation served as an effective tool of research for many decades. Jack Major (1951) extended it to embrace the entire ecosystem – soils, vegetation, and animal life. Jenny (1961, 1980) offered an extension that included ecosystems (entire sections of landscapes). He derived a general state-factor equation of the form:

$$l, s, v, a = f(L_0, P_x, t)$$

where *l* is ecosystem properties such as total carbon content, primary production, and respiration; *v* is vegetation properties such as biomass, species frequency, and sodium content; *a* is animal properties such as size, growth rate, and colour; and *s* is soil properties such as pH, texture, humus content; L_0 is the initial state of the system, that is, the assemblage of properties at time zero when development starts (the *L* stands for the ecosystem, or *l*arger system, of which the soil is part); P_x are external flux potentials; and *t* is the age of the system. The state factors are groups of variables associated with L_0 and P_x. Parent material, *p*, and the original topography and water table, *r*, define the initial state of the system. The external flux potentials are environmental properties that lead to additions and subtractions of matter and **energy** to and from the system. They include environmental climate, *cl*, and a biotic factor, *o*, comprising fauna and flora as a pool of species or genes, active or dormant, that happen to be in the ecosystem at time zero or that enter it later. The biotic factor is thus distinct from the vegetation that grows as the system develops; this appears as a system property on the left-hand side of the equation. Other external fluxes would include dust storms, floods, and the additions of fertilizers. In an extended form, the general state-factor equation is:

$$l, s, v, a = f(cl, o, r, p, t, \ldots)$$

This brings us back to the 'clorpt' equation, only this time it applies to ecosystems and not just soils. The latest version of the 'clorpt' equation considers the place of the human species in the state-factor theory of ecosystems (Amundson and Jenny 1991). Jenny's general state-factor equation applied to an ecosystem includes animals and plants as system properties, as well as soil. It provides a useful way of thinking about relations between living things and their **environment**. It is exceedingly valuable, especially as a conceptual tool. Some pedologists and ecologists still cite it and use it, often in their own variant forms.

Further reading: Schaetzl and Anderson 2005

GAIA HYPOTHESIS

The Gaia hypothesis asserts that, shortly after it first appeared, life has been at the helm, exercising near total homeostatic control of the terrestrial **environment**. To be sure, life wields a potent influence on the composition of the atmosphere, producing a chemical disequilibrium, as seen in the high concentration of reactive atmospheric oxygen. Photosynthesis sustains this chemical disequilibrium by releasing oxygen and removing carbon dioxide from the air. It occurred to James Lovelock (1965), an atmospheric chemist, that such a non-equilibrium atmospheric state would be a guide to the presence of life on other planets. This line of thought led Lovelock, in collaboration with microbiologist Lynn Margulis, to design the Gaia hypothesis (Lovelock 1972, 1979, 1988; Lovelock and Margulis 1974), the novelist William Golding suggesting the name. A key component of the Gaia hypothesis is the assertion that the biosphere maintains atmospheric **homeostasis**, primarily though negative **feedback** processes, and in so doing sustains environmental conditions conducive to life. This simple idea has proved extremely controversial and has stimulated scientific debate.

The central premise of the Gaia hypothesis comes in two versions, which give rise to the strong Gaia hypothesis and the weak Gaia hypothesis (Kirchner 1991). In the strong Gaia hypothesis, the biosphere is able to change the environment to suit life; in the weak Gaia hypothesis, the biosphere is able hold the environment within limits fit for life. Weak Gaia predicts that life wields a substantial influence over some features of the abiotic world, mainly by playing a pivotal role in **biogeochemical cycles**. Life's influence is sufficient to have produced highly anomalous environmental conditions in comparison with the flanking terrestrial planets, Venus and Mars. Notable anomalies include the presence of highly reactive gases (including oxygen, hydrogen, and methane) coexisting for long times in the atmosphere, the stability of the Earth's temperature in the face of increasing solar luminosity, and the relative alkalinity of the oceans. By interacting with the surface materials of the planet, life has sustained these unusual conditions of temperature, chemical composition, and alkalinity for much of geological time. The weak Gaia hypothesis does not call upon anything other than mechanistic processes to explain terrestrial **evolution**, but it does contend that the biosphere built and maintains the abiotic portion of the ecosphere. Strong Gaia is, to some, the unashamedly teleological idea that the Earth is a superorganism controlling the terrestrial environment to suit its own ends. In his earlier writings, Lovelock seemed to favour strong Gaia. He believed that it is useful to regard the planet Earth, not as an inanimate globe of rock, liquid, and gas driven by geological processes, but as a sort of biological superorganism, a single life-form, a living planetary body that adjusts and regulates the conditions in its surroundings to suit its needs (e.g. Lovelock 1991). In a recent book, Lovelock (2000) stepped back a little from his original 'somewhat outrageous statements'. He explained that, to make himself heard, he had to act like a neglected child who behaves badly in order to gain attention, and simply used the metaphor of a living Earth to make humourless biologists think that he really thought the Earth was alive and reproduces, whereas in fact he did not.

Gaian scientists claim that traditional biology and geology offer ineffective methods with which to study the planetary organism. The right tool for the job, they contend, is geophysiology – the science of bodily processes writ large and applied to the entire planet, or at least that outer shell encompassing the biosphere. The differences of approach and emphasis are fundamental – if the strong Gaia hypothesis should be correct, and the Earth really is an integrated superorganism, then the biosphere will regulate and maintain itself through a complex system of homeostatic mechanisms, just as the human body adjusts to the vicissitudes of its surroundings. Consequently, the biosphere may be a far more robust and resilient beast than has often been suggested. For instance, homeostatic mechanisms may exist for healing the hole in the ozone layer or preventing the global thermometer from blowing its top.

To Lovelock (1991), the Gaia hypothesis, in all forms, suggests three important things. First life is a global, not local, phenomenon. It is not possible for sparse life to inhabit a planet – there must be a global film of living things because organisms must regulate the conditions on their planet to overcome the ineluctable forces of physical and chemical evolution that would render it uninhabitable. Second, the Gaia hypothesis adds to Darwin's vision by negating the need to separate species evolution from environmental evolution. The evolutions of the living and non-living worlds are so tightly knit as to be a single indivisible process. A coherent coupling between organisms and the material environment, and not just survival of the fittest, is a measure of evolutionary success. Third, the Gaia hypothesis may provide a way to view the planet in mathematical terms that 'joyfully accepts the nonlinearity of nature without being overwhelmed by the limitations imposed by the chaos of complex dynamics' (Lovelock 1991, 10).

A major drawback with the Gaia hypothesis is its unsuitability for testing. In an attempt to overcome this problem, Axel Kleidon (2002) tried to express the Gaia hypothesis in terms that enable the formulation of testable null hypotheses. He used gross primary production (GPP), which is the global gross uptake of carbon by organisms, to describe biotic activity. GPP seems a fair measure of how beneficial environmental conditions are for life – the more favourable the conditions, the higher the GPP. With this definition, he formulated a set of hypotheses focusing on how GPP for an environment including biotic effects compared to the hypothetical value of GPP for an environment without biotic effects – in effect, a planet with life compared to a lifeless planet. His approach did not focus on whether particular biotic feedbacks are positive or negative, but on the *sum* of all biotic effects. It would be difficult to construct the environmental conditions without biotic effects in the real world, but numerical simulation models provide a means to do so. Using climate model simulations of extreme vegetation conditions – a Desert World and a Green World (called a Green Planet by Kleidon) – Kleidon (2002) showed that, overall, terrestrial vegetation generally leads to a climate that is more favourable to carbon uptake. He concluded that 'life has a strong tendency to affect Earth in a way which enhances the overall benefit (that is, carbon uptake)' (Kleidon 2002). This interesting line of research has its critics and the debate is certainly ongoing (e.g. Lovelock 2003; Kleidon 2004, 2007; Volk 2007; Phillips 2008), but it cannot be denied that the Gaia hypothesis is a thought-provoking idea that has generated much very penetrating discussion.

Further reading: Lovelock 1988

GENERAL CIRCULATION OF THE ATMOSPHERE

Air moves relentlessly and atmospheric motion is the key to understanding Earth's climates and so to understanding many Earth surface patterns and processes. On a planetary scale, the temperature gradient between the equator and the poles is the foremost driver of air movements. This thermal gradient drives a vast overturning of air, first suggested by Edmund Halley in 1686 and elaborated by George Hadley in 1735, called the Hadley circulation. On Venus, there is one grand Hadley cell in each hemisphere: a huge convective current of air rises at the Venusian equator, moves polewards, sinks over the poles, and then returns at ground level to its equatorial origin. On the Earth, the Hadley circulation breaks down into three component cells in each hemisphere (Figure 16). Heat released as water evaporated from the tropical oceans condenses, mainly in the inter-tropical convergence zone (ITCZ) or equatorial low-pressure trough, and largely powers the equatorial Hadley cell. At ground level, the air returning towards the ITCZ produces the trade winds. The middle or Ferrel cell flows in the reverse direction, that is, equatorwards aloft and polewards at the surface. The third cell, known as the polar cell, is rather weak.

Because the Earth rotates, winds in the tropics tend to blow north-east to south-west (Northern Hemisphere) or south-east to north-west (Southern Hemisphere). In middle and high latitudes, where the

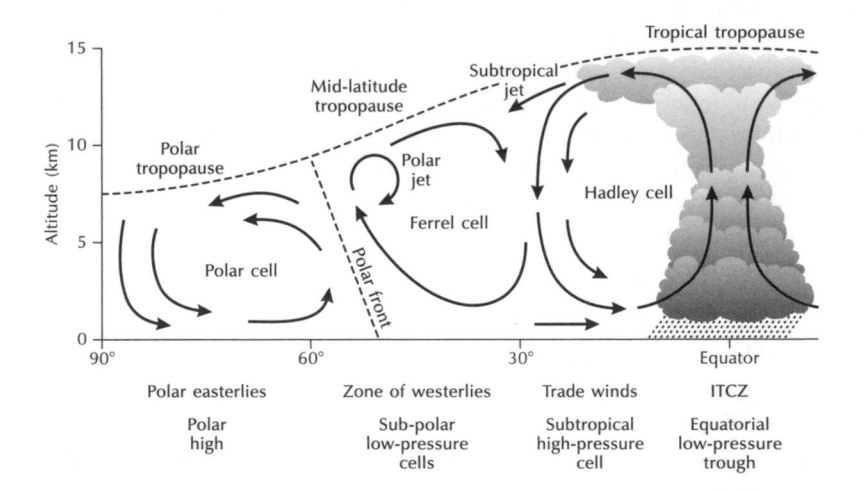

Figure 16 The general circulation of the atmosphere.

Source: After Huggett (2007c)

Coriolis force is strongest, they have a strong east-to-west component. In addition, easterly winds predominate in the upper troposphere in the tropics, and westerly winds elsewhere. The upper air westerlies are the main flow of the atmosphere. They form a circumpolar vortex, the fastest flowing ribbons of which are jet streams. This vortex progresses from shallow zonal waves rushing west to east to great meandering meridional loops. The full sequence of the 'index cycle', so named because the zonal index (the average hemispherical pressure gradient between 35° and 55° N) runs from low to high and back again over a four-to-six-week period (Namias 1950). The tropical easterlies may also form jet streams, especially in summer.

The tripartite Hadley circulation has a smaller 'eddy' circulation superimposed upon it. Baroclinic instabilities and the development of standing waves create this eddy circulation. Baroclinic instabilities arise when steep temperature and pressure gradients cross one another. The result is that baroclinic waves develop, forming the disturbances known as the travelling cyclones and anticyclones characteristic of mid-latitudes. These disturbances play a major role in transferring heat to the upper troposphere and towards the poles. In tropical latitudes, large and intense wave cyclones are absent because of the weak Coriolis force. Instead, disturbances take the form of easterly waves, which are slow moving troughs of low pressure within the trade wind belt (Dunn 1940; Riehl 1954); weak equatorial lows, which form at the heart of equatorial troughs and serve as foci for individual convective storms along the ITCZ; and powerful tropical cyclones.

The latitudinal temperature gradient and the general circulation serve to define broad hot, cold, wet, and dry zones. However, seasonal climatic features, particularly those associated with tropical cyclones and the poleward swing of the ITCZ in the summer hemisphere and associated monsoon circulations, are enormously important in defining the climates of many places, including some of the densely populated ones. The seasonal shift of the ITCZ in the summer hemisphere explains such seasonal rainfall patterns as summer rains in savannah zones, winter rains in west coast Mediterranean climates, summer rains in the east coast (China) climates and in the tropical margin zones like India and northern Australia. Monsoon circulations develop in parts of the tropics and are connected with the northward and southward swing of the ITCZ. In southern Asia, monsoons involve a changeover from north-east winds in winter to south-west monsoon winds in summer.

Tropical cyclones – called hurricanes in the Atlantic Ocean, typhoons in the Indian and Pacific Oceans, and willy-willies in

Australia – are large-scale, though seasonal, features of the general circulation. They begin as weak low-pressure cells over very warm ocean water (27°C and more) between 8° and 15° N and S that grow into deep circular lows. Tropical cyclones tend to define the character of such places as Taiwan, Luzon, several Pacific Islands, the Bay of Bengal, Florida, the Caribbean, and north-western and north-eastern Australia. Paradoxically, while tropical cyclones may cause severe coastal damage, witness the flooding of New Orleans resulting from hurricane Katrina in September 2005, they can be the major rain-producing systems of the year some areas, including much of inland Australia.

The generation of stationary eddies (standing waves) by massive mountain ranges and plateaux, and by temperature differences between oceans and continents, influences atmospheric circulation patterns. The Rockies, for example, tend to anchor the westerly jet stream over North America. Similarly, the Tibet plateau affects the position of the jet stream, which in turn influences the invasion of the south-west monsoon, which occurs as the jet stream shifts from the south or southern edge of the Tibet plateau.

Seasonal temperature changes result partly from differences in thermal properties of land and sea. A landmass will have hotter summers and colder winters than an ocean occupying the same latitude. This is the effect of continentality. It arises because land has two to three times the heat capacity of sea, and because heat is conducted downwards more slowly from the land surface than it is by turbulent mixing from the ocean surface. For these reasons, the annual and diurnal ranges of surface and air temperatures are much larger in continental climates than in oceanic climates. The influence of oceans extends over continental areas lying next to oceans, especially non-mountainous areas on western seaboards where maritime air masses move inland. Various indices describe the degree of continentality. The Northern Hemisphere displays more marked effects of continentality than does the Southern Hemisphere, simply because the Northern Hemisphere contains about twice as much land as the Southern Hemisphere.

Smaller circulations of air include individual eddies revealed by the swirling of autumn leaves, thunderstorms, and land and sea breezes. These ride within the larger-scale hurricanes, fronts, cyclones and anticyclones, and hemispheric and global circulations.

Further reading: Barry and Chorley 2003

GENERAL CIRCULATION OF THE OCEANS

Ocean water, like air in the atmosphere, circulates. The dominant paths of flow are ocean currents. Prevailing surface winds drive nearly all large currents at the top of the ocean. The main feature of the surface-ocean circulation is a set of gyres (Figure 17a). These are slow-moving vortices that occur in the oceans of the Northern and Southern Hemispheres. They rotate clockwise in the Northern Hemisphere and anticlockwise in the Southern Hemisphere, tracking the passage of air around the subtropical high-pressure cells. A large gyre, or circumpolar current, also moves around Antarctica. Near the equator, ocean currents travel westwards. On approaching land, they are diverted to the north and south to form warm currents that run parallel to the coastline. In the Pacific Ocean, some of the water returns eastwards, approximately along the equator, as an equatorial counter-current. Once in middle latitudes, the warm currents are forced eastwards by the westerly winds. They travel across the oceans until they reach the western edges of continents, which they keep relatively warm in winter. They then join either the Arctic or the Antarctic circulation, or else they return towards the equator as cool currents. They are often associated with cold-water upwellings along continental margins, as with the Humboldt Current that runs down the coastline of Peru and Chile. The presence of cold currents in tropical waters and warm currents in temperate and polar waters can affect regional and even global climates. The warm North Atlantic Current (North Atlantic Drift) moderates winter temperatures in the British Isles and northwest Europe. The cold Humboldt Current off western Ecuador, Peru, and Chile reduces temperatures and increases **aridity** in those countries, and especially in Chile. Periodically, the water in the current becomes warmer as part of the El Niño–Southern Oscillation, which can affect climates around the world (see **tele-connections**).

The ocean deep-water circulation is somewhat different from the surface currents (Figure 17b). In the ocean deeps, a salty current threads its way around the world's ocean basins (Broecker and Denton 1990). It starts in the North Atlantic Ocean where northwards-flowing warm and normally salty water is chilled by cold Arctic air and by evaporation. This already relatively dense water then sinks to the ocean bottom and flows southwards, through the North Atlantic and South Atlantic Oceans. It is warmer and less dense than the frigid surface waters off Antarctica, so it rises to the surface, where it is

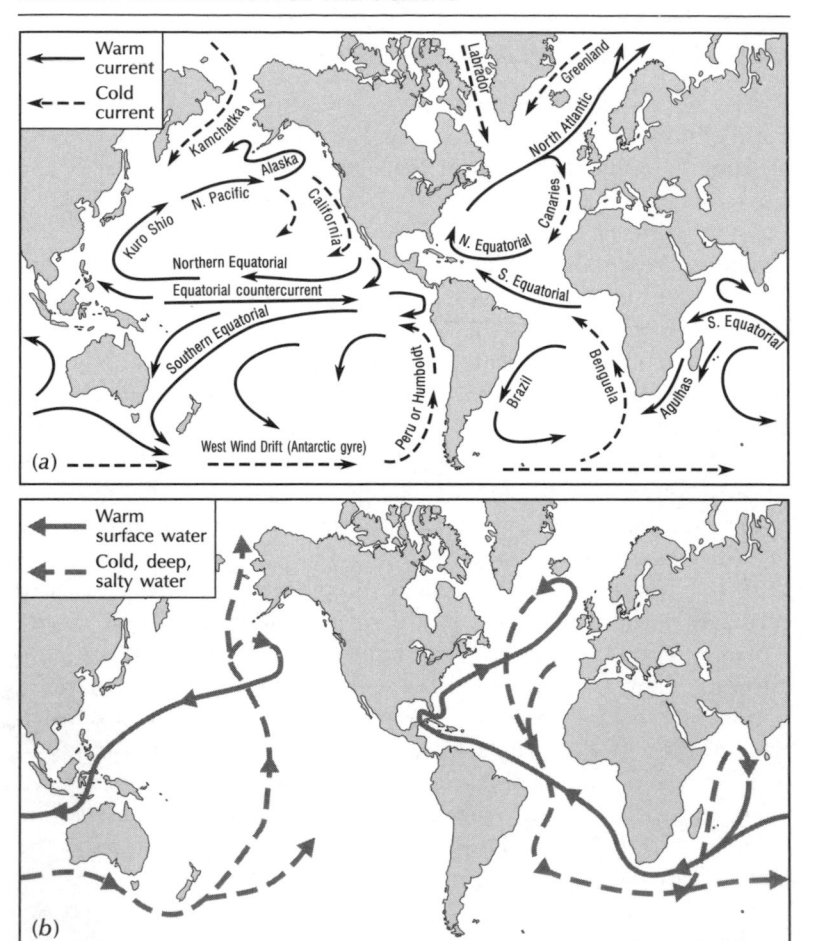

Figure 17 Ocean currents. (a) Global surface ocean currents. (b) Major deep water and surface currents between oceans.

chilled again and plunges back to the ocean depths. It then turns east-wards between South Africa and Antarctica. One branch moves into the Indian Ocean and another continues eastwards, passing by Australia and New Zealand, and northwards into the Pacific Ocean. In the Indian and Pacific Oceans, the northward flow of cold bottom waters is compensated by a southward flow of surface waters in the gyres. Counterflow of warm surface water in the Atlantic Ocean is rapidly caught up in the strong southwards current of the deep, cold water. Disruption of this 'global conveyor' system can cause severe and sudden climatic change (e.g. Broecker 1995; Rahmstorf 2003; Lenton *et al.* 2008).

GEOCHRONOLOGY

Geochronology is the science of determining the absolute age of rocks, fossils, landforms, and sediments within a certain degree of uncertainty inherent within the method used. Geochronologists have at their disposal a broad range of methods for dating events in Earth history. Some are more precise than others are. Four categories are recognized: numerical-age methods, calibrated-age methods, relative-age methods, and correlated-age methods. Numerical-age methods produce results on a ratio (or absolute) timescale, pinpointing the times when **environmental change** occurred. This information is crucial to a deep appreciation of environmental change: without dates, nothing much of use can be said about rates. Calibrated-age methods may provide approximate numerical ages. Some of these methods are refined and enable age categories to be assigned to deposits by measuring changes since deposition in such environmental factors as **soil** genesis or rock weathering. Relative-age methods furnish an age sequence, simply putting events in the correct order. They assemble the 'pages of Earth history' in a numerical sequence. The Rosetta-stone of relative-age methods is the principle of stratigraphic superposition. This states that, in undeformed sedimentary sequences, the lower strata are older than the upper strata. Some kind of marker must be used to match stratigraphic sequences from different places. Traditionally, geologists use fossils for this purpose. Distinctive fossils or fossil assemblages can be correlated between regions by identifying strata that were laid down contemporaneously. This was how such celebrated geologists as William ('Strata') Smith (1769–1839) first erected the stratigraphic column. Although this technique was remarkably successful in establishing the broad development of Phanerozoic sedimentary rocks, and rested on the sound principle of superposition, it is best used in partnership with numerical-age methods. Used conjointly, relative-age methods and numerical-age methods have helped to establish and calibrate the geological timetable.

Correlated-age methods do not directly measure age, but suggest ages by showing equivalence to independently dated deposits and events. There are six groups of dating techniques: sidereal, isotopic, radiogenic, chemical and biological, geomorphic, and correlation (Colman and Pierce 2000). As a rule, sidereal, isotopic, and radiogenic methods give numerical ages, chemical and biological and geomorphic methods give calibrated or relative ages, and correlation methods give correlated ages. However, some methods defy such ready classification. For instance, measurements of amino-acid racemization may yield results as relative age, calibrated age, correlated age, or numerical age, depending on the extent to which calibration and control of environmental variables constrain the reaction rates. Another complication is that, although isotopic and radiogenic methods normally produce numerical ages, some of them are experimental or empirical and need calibration to produce numerical ages.

Further reading: Colman and Pierce 2000

GEODIVERSITY

Geodiversity or geological diversity is the diversity of minerals, rocks, fossils, **soils**, landforms, and associated processes that constitute Earth's **topography**, landscapes, and underlying geological structures. The range of geological and palaeontological features in a region determines its geodiversity. High geodiversity occurs in areas with many different geological structures, especially if these belong to differing geological periods (such as in the British Isles). Low geodiversity tends to occur in areas underlain by large tracts of similar geological structures.

The concept of geodiversity, although far less well known than the concept of **biodiversity**, is starting to develop its own momentum. A growing number of Earth scientists see the conservation of geodiversity as imperative in its own right, and as an essential adjunct to biodiversity and cultural conservation programmes and as a key element of people's natural heritage. The argument runs that geodiversity has many values – aesthetic, intrinsic, ecological, economic, scientific, heritage, educational – and merits protection. Problems of conserving geodiversity, the subject matter of geoconservation, have only recently surfaced. Traditionally, Earth scientists thought many geological and geomorphic features tough enough not to require protection or management. However, current thinking is turning to concerns over a loss of geodiversity.

Geological and geomorphic features are vulnerable to damage and destruction in several ways. Developing land for urban or industrial use can damage natural topography or geological exposures. The infilling of quarries can bury exposures of scientific or educational interest. Natural weathering, the growth of vegetation, and trampling by livestock may damage or obscure geological features. Engineering works to rivers and coasts may interfere with natural processes and damage natural features. To avoid such damage, many local authorities and other bodies in the United Kingdom have developed Geodiversity Action Plans. A case in point is the North Pennines Area of Outstanding Natural Beauty (AONB) Partnership (an alliance of twenty-two statutory agencies, local authorities, and voluntary or community organizations). In close cooperation with the British Geological Survey, this body developed a Geodiversity Action Plan for the North Pennines, the first UK-protected landscape to have such a plan. In June 2003, the North Pennines AONB became Britain's first European Geopark. Geoparks are areas with outstanding geology, where there is a clear strategy and ongoing action for the conservation of its Earth heritage and promoting the enjoyment and understanding of geology. In February 2004, UNESCO announced the establishment of a new Global Geoparks Network, with the North Pennines AONB Partnership being amongst the founding members of this new worldwide family of special places (www.unesco.org/science/earth/geoparks.shtml).

Further reading: Gray 2003

GEOGRAPHICAL CYCLE

The 'geographical cycle' (cycle of erosion, geomorphic cycle), expounded by William Morris Davis, was the first modern theory of landscape **evolution** (e.g. Davis 1899, 1909). Although it no longer has much currency and other theories have superseded it, the geographical cycle was a key concept in physical geography for many decades. Its appeal seems to have lain in its theoretical tenor and in its simplicity (Chorley 1965). It was quite revolutionary and outstanding in its time, helping to modernize physical geography, to create the field of geomorphology, and to spawn the once highly influential field of denudation chronology.

Davis's geographical cycle starts with assumed rapid uplift. Geomorphic processes, without further complications from tectonic movements, then gradually wear down the raw **topography**. Furthermore, slopes within landscapes decline through time – maximum slope angles slowly lessen (though few field studies have substantiated this claim). In consequence, topography reduces, little by little, to an extensive flat region close to base level – a peneplain – with occasional hills called monadnocks after Mount Monadnock in New Hampshire, USA, which are local erosional remnants, standing conspicuously above the general level. The reduction process creates a time sequence of landforms progressing through the stages of youth, maturity, and old age. Davis originally designed his cycle to account for the development of humid temperate landforms produced by prolonged wearing down of uplifted rocks offering uniform resistance to erosion. He and others extended it to different landforms, including arid landscapes, glacial landscapes, and periglacial landscapes, to landforms produced by shore processes, and to karst landscapes.

Davis's theory of geographical cycles has flaws. In particular, real landscapes are not quite as orderly as Davis envisaged and erosion occurs during the uplift process. In addition, the use of the terms youth, maturity, and old age, which Davis borrowed from biology, are misleading and much censured (e.g. Ollier 1967; Ollier and Pain 1996, 204–5).

GEOLOGICAL CYCLE

The geological cycle is the repeated creation and destruction of crustal material – rocks and minerals (Figure 18). Volcanoes, folding, faulting, and uplift all bring igneous and other rocks, water, and gases to the base of the atmosphere and hydrosphere. Once exposed to the air and

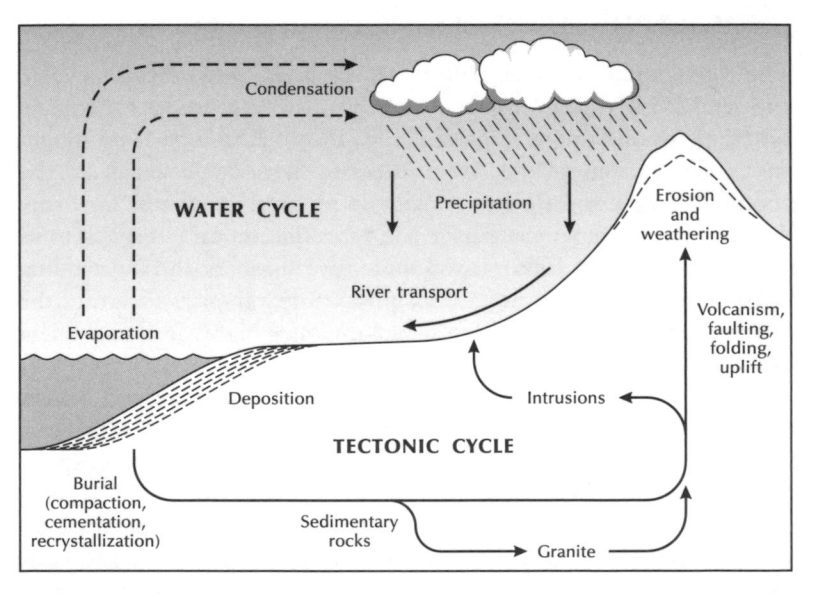

Figure 18 The rock cycle, the water cycle, and their interaction.
Source: After Huggett (2007a)

meteoric water, these rocks begin to decompose and disintegrate by the action of weathering. Gravity, wind, and water transport the weathering products to the oceans. Deposition occurs on the ocean floor. Burial of the loose sediments leads to compaction, cementation, and recrystallization and so to the formation of sedimentary rocks. Deep burial may convert sedimentary rocks into metamorphic rocks. Other deep-seated processes may produce granite. If uplifted, intruded or extruded, and exposed at the land surface, the loose sediments, consolidated sediments, metamorphic rocks, and granite may join in the next round of the rock cycle.

Volcanic action, folding, faulting, and uplift may all impart potential **energy** to the toposphere, creating the 'raw relief' on which geomorphic agents may act to fashion the marvellously multifarious array of landforms found on the Earth's surface – the physical toposphere. Geomorphic or **exogenic forces** or agents are wind, water, waves, and ice, which act from outside or above the toposphere; these contrast with endogenic forces or agents (tectonic and volcanic), which act upon the toposphere from inside the planet.

The surface phase, and particularly the land-surface phase, of the rock cycle is the domain of geomorphologists. The flux of materials across the land surface is, overall, unidirectional and is a cascade rather than a cycle. The basics of the land–surface debris cascade are as follows. Weathering agents move into the **soil** and rock along a weathering front, and in doing so fresh rock is brought into the system. Material may be added to the land surface by deposition, having been borne by wind, water, ice, or animals. All the materials in the system are subject to transformations by the complex processes of weathering. Some weathering products revert to a rock-like state by further transformations: under the right conditions, some chemicals precipitate out from solution to form hardpans and crusts. And many organisms produce resistant organic and inorganic materials to shield or to support their bodies. The weathered mantle may remain in place or it may move downhill. It may creep, slide, slump, or flow downhill under the influence of gravity (mass movements) or moving water may wash or carry it downhill. In addition, the wind may erode it and take it elsewhere.

GLOBAL WARMING

There is growing evidence that the human population may be having a global climatic impact, triggering a warming of the atmosphere and oceans largely through increasing the concentration of atmospheric greenhouse gases. However, natural cycles of **climate change** may also lead to fluctuations in atmospheric carbon dioxide levels and in the levels of some other greenhouse gases, which may bring about a warming and cooling of the atmosphere. Natural climate change must provide a backdrop against which to assess the extent of human-induced global warming. Indeed, current global warming may have a natural component related to orbital forcing that would keep temperatures rising even without an increase in greenhouse gases (Kukla and Gavin 2004), although a forceful human signal now seems incontrovertible (e.g. Foukal *et al.* 2006).

Evidence is mounting that the Earth is warming. Strong observational indicators include the following. First, the global average surface temperature (the average of near-surface air temperature over land, and sea-surface temperature) has increased since 1861 (Figure 19a). During the twentieth century, the increase was about 0.6°C, although the warming was not even, most of it taking palace in the periods 1910 to 1940, and 1976 to 2000. Second, globally, it is likely the 1990s were the warmest decade and 1998 the warmest year in the instrumental record since 1861. Third, twentieth-century warming in the Northern Hemisphere is likely to have been the fastest for any time over the last 1,000 years; the 1990s were the warmest decade and 1998 the warmest year (Figure 19b).

Almost without exception, climate models that simulate the effect of much higher burdens of carbon dioxide and trace gases in the atmosphere predict that the Earth will be a warmer and more humid planet during the current century. Predictions suggest a globally averaged surface temperature in the range 1.4–5.8°C over the period 1990 to 2100, depending upon the scenario used (Figure 20). The basic reasons for the increased temperature and humidity are not difficult to grasp. The higher the concentration of greenhouse gases in the atmosphere, the greater the amount of infrared radiation emitted from the Earth's surface absorbed by the atmosphere, and so the hotter the atmosphere. With a warmer atmosphere, evaporation of water from the world's oceans increases, so leading to a more vigorous pumping of water round the **hydrological cycle**. This results in an increased occurrence of droughts and of very wet conditions, the last created by deeper thunderstorms with greater rainfall. Tropical cyclones also become more destructive. The increased humidity of the air may itself boost greenhouse warming since water vapour absorbs infrared radiation.

Interestingly, climate models predict an uneven warming of the atmosphere, the land areas warming faster than the seas, and particularly the land areas at northern high latitudes during the cold season. Predictions indicate that the northern regions of North America and

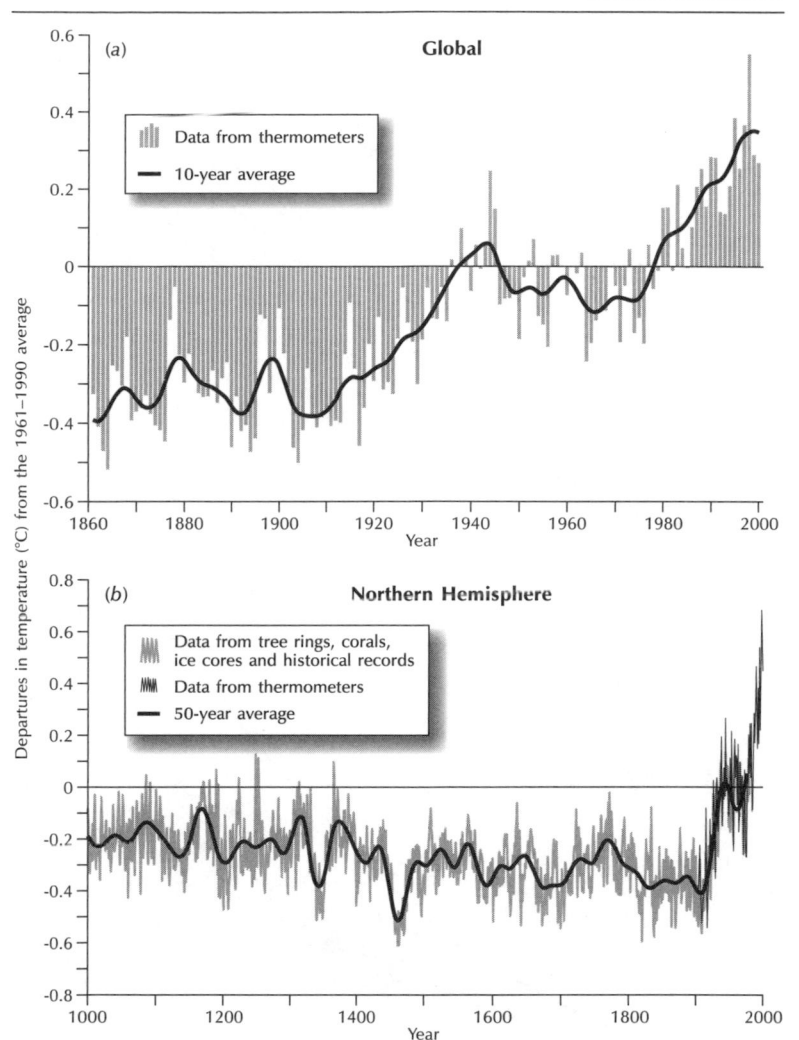

Figure 19 Variations of the Earth's surface temperature, (a) globally over the years 1860–2000; (b) in the Northern Hemisphere over the last millennium.

Source: Adapted from Houghton *et al.* (2001)

northern and central Asia will warm 40 per cent more than the global average. Conversely, south and south-east Asia in summer and southern South America in winter will warm less than the global mean. It also seems likely that surface temperatures will become more El Niño-like in the tropical Pacific Ocean, with the eastern tropical Pacific warming

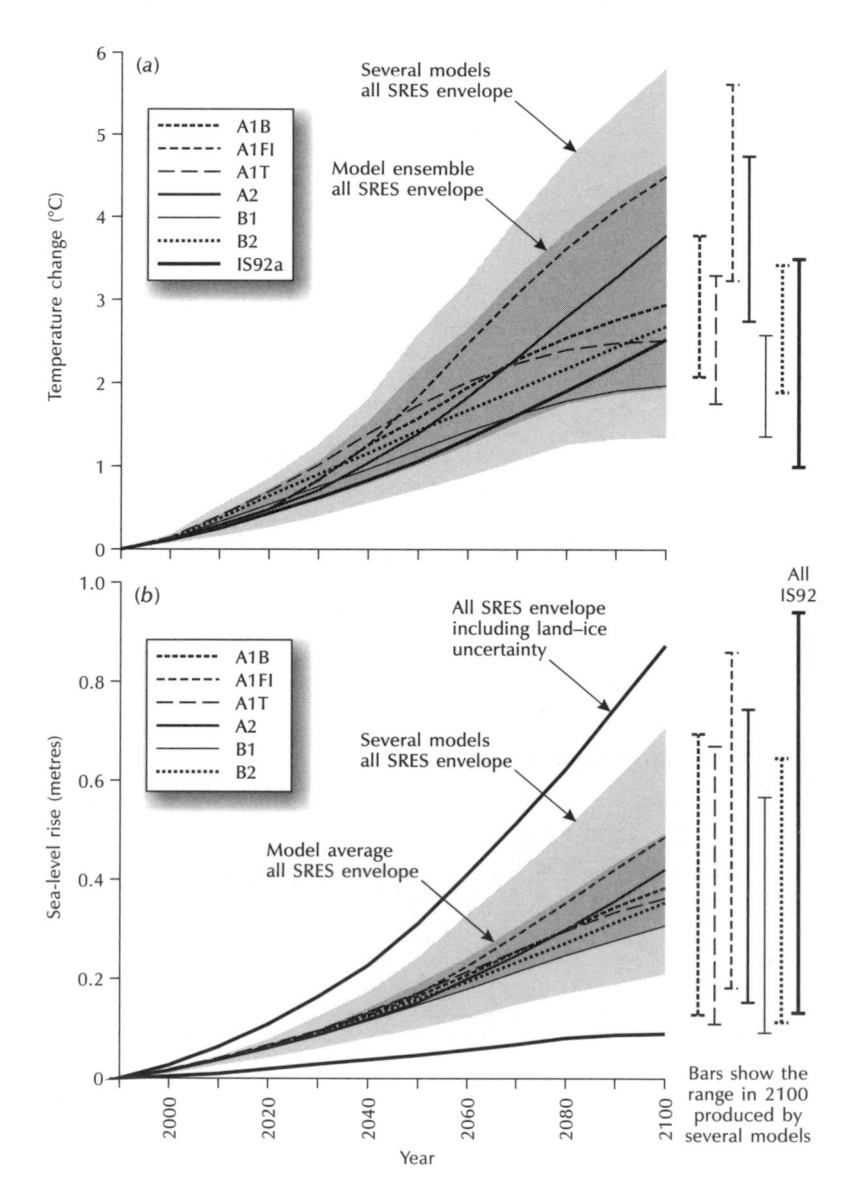

Figure 20 Predictions of climatic change during the twenty-first century.
(a) Temperature changes. (b) Sea-level rise. Scenario A1F1 (fossil-fuel intensive) is the worst-case scenario; scenario B1 (the 'green scenario') is the best-case scenario.

Source: Adapted from Houghton *et al.* (2001)

more than the western tropical Pacific. This differential warming should substantially alter the global pattern of evaporation and precipitation and cause radical changes of climate in most regions outside the tropical zone. The models suggest that the global average water-vapour concentration and precipitation will rise over the twenty-first century. After 2050, northern mid- to high-latitudes and Antarctica will have wetter winters. At low latitudes, some regions will become wetter and some drier. Large year-to-year variations in precipitation are likely in most regions with higher predicted precipitation (see Table 5).

Further reading: Houghton 2004; Lomborg 2007

Table 5 Estimated confidence in observed and predicted extreme weather and climate events

Phenomenon	Confidence in observed changes (second half of twentieth century) (%)	Confidence in predicted changes (twenty-first century) (%)
Higher maximum temperatures and more hot days in nearly all land areas	66–90	90–99
Higher minimum temperatures, fewer cold days and frost days overly near all land areas	90–99	90–99
Reduced daily temperature range over most land areas	90–99	90–99
Increase of heat index[a] over land areas	66–90, over many areas	90–99, over most areas
More intense precipitation events	66–99, over many Northern Hemisphere mid- and high-latitude land areas	90–99, over most areas
Increased summer continental drying and associated risk of drought	66–99, in a few areas	66–90, over most mid-latitude continental interiors (other areas lack consistent predictions)
Increase in tropical cyclone peak-wind intensities	Not observed in the few analyses available	66–90, over some areas
Increase in tropical cyclone mean and peak precipitation intensities	Insufficient data for assessment	66–90, over some areas

[a]The heat index combines temperature and humidity as a measure of their effects on human comfort.

Source: Adapted from Houghton *et al.* (2001)

GRADUALISM

Gradualism is a school of thought arguing that, throughout the history of the Earth, geological and biological processes have operated at rates more or less the same as those observed in the present day. Most commentators credit James Hutton (1726–97) as the father of gradualism. Admittedly, well before Hutton, Aristotle and Leonardo da Vinci discussed the effectiveness of geological agencies – wind, rain, sea, sun, and earthquakes – in refashioning the Earth's surface, but Hutton was assuredly the architect of the first full-blown gradualist system of Earth history. He saw the world as a perfect machine that would run forever through its cycles of decay and repair – crustal uplift, erosion, transport, deposition, compaction and consolidation, and renewed uplift – now called the geological or sedimentary cycle. John Playfair (1748–1819) energetically championed Hutton's revolutionary ideas; Charles Lyell (1797–1875), in his celebrated *Principles of Geology* (1830–33), embellished and elaborated upon them. Lyell, the arch gradualist, carefully and convincingly argued that the slow and steady operation of present geological processes could explain the apparently enormous changes that the Earth had evidently suffered in the past. Gradualism was an essential ingredient of Lyell's **uniformitarianism** that pervaded thinking in the Earth and life sciences until **catastrophism** made its recent comeback.

In biology, evolutionists with gradualist convictions opine that life evolves steadily, little by little, in a stately fashion. The notion of gradual change in the organic world occurred to many pre-Darwinian thinkers, including Benoît de Maillet, Georges Louis Leclerc, Comte de Buffon, Erasmus Darwin, Jean-Baptiste Pierre Antoine de Monet de Lamarck, Robert Chambers, and Bernhard von Cotta (for citations, see Huggett 1990, 1997b). Charles Robert Darwin was the first person to arrive at the view that animals and plants might evolve gradually, in a definite direction, owing to external influences acting on small and random variations. Darwin's dictum that *Natura non facit saltum* (Nature does not make jumps) is a catchphrase for the gradualistic school of evolutionary change. Neo-Darwinians are micromutationists, subscribing to the view that **evolution** proceeds by the gradual accumulation of small genetic changes. However, the gradualism of extreme micromutationism is probably too slow to account for, and seems inconsistent with, the observed changes in the fossil record. An influential group of micromutationists, which includes among its number George Gaylord Simpson and Ernst Mayr, allows a reorganization of the genotype within relatively few generations in a small colony of organisms. In addition, it sees such periods of relatively fast genetic change as a possible seat of bigger evolutionary changes, including the origination of major groups such as the mammals and the angiosperms. This notion of relatively rapid **speciation** shifts the emphasis away from gradual changes, in the strict sense employed by Darwin, towards punctuationalism (see **catastrophism**).

Further reading: Huggett 1997b

HABITAT

Individuals, species, and **populations**, both marine and terrestrial, tend to live in particular places. These places are **habitats**. A specific set of environmental conditions – radiation and light, temperature, moisture, wind, fire frequency and intensity, gravity, salinity, currents, **topography**, **soil**, substrate, geomorphology, human **disturbance**, and so forth – characterizes each habitat.

Habitats come in all shapes and sizes, occupying the full sweep of geographical scales. They range from small, through medium and large, to very large. Small-scale or microhabitats are a few square centimetres to at very most a square kilometre in area. They include leaves, the soil, lake bottoms, sandy beaches, talus slopes, walls, riverbanks, and paths. Medium-scale or mesohabitats have areas up to about 10,000 km^2; that is, a 100×100 kilometre square, which is about the size of Cheshire, England. Similar features of geomorphology and soils, a similar set of disturbance regimes, and the same regional climate influence each main mesohabitat. Deciduous woodland, caves, and streams are examples. Large-scale or macrohabitats have areas up to about 1,000,000 km^2, which is about the size of Ireland. Very large-scale or megahabitats are regions more than 1,000,000 km^2 in extent and include continents and the entire land surface of the Earth.

It is probably true to say that no two species have exactly the same living requirements. There are two extreme cases – fussy species or habitat specialists and unfussy species or habitat generalists – and all grades of 'fussiness' between. Habitat specialists have very precise living requirements. In southern England, the red ant, *Myrmica sabuleti*, needs dry heathland with a warm south-facing aspect that contains more than 50 per cent grass species, and that has been disturbed within the previous five years (Webb and Thomas 1994). Other species are less finicky and thrive over a wider range of environmental conditions. The three-toed woodpecker (*Picoides tridactylus*) lives in a broad swath of cool temperate forest encircling the Northern Hemisphere. Habitat generalists manage to eke out a living in a great array of environments. In the plant kingdom, the broad-leaved plantain (*Plantago major*), typically a species of grassland habitats, is found almost everywhere except Antarctica and the dry parts of North Africa and the Middle East. In the British Isles, it seems indifferent to climate and soil conditions, growing in all grasslands on acid and alkaline soils alike. It also lives on paths, tracks, disturbed habitats (spoil heaps, demolition sites, arable land), pasture and meadows, road verges, riverbanks, mires, skeletal habitats, and as a weed in lawns and sports fields. In woodland, it lives only in relatively unshaded areas along rides, and it does not live in aquatic habitats or tall herb communities. Despite these clear examples, the distinctions between habitat generalists and specialists are not always so definite, and sometimes they can seem almost arbitrary. In birds, for instance, some researchers take specialists as species breeding in one habitat type and generalists as species breeding in two or more habitat types (e.g. Skórka *et al.* 2006).

HABITAT LOSS AND HABITAT FRAGMENTATION

Habitat loss and habitat fragmentation pose severe threats to global **biodiversity**, having an adverse impact on nearly all taxonomic groups, including plants, invertebrates, amphibians, reptiles, birds, and mammals. In consequence, they have emerged as a foremost research theme in conservation biology (Haila 2002; Fazey *et al.* 2005). However, the lack of a clear conceptual foundation and the loose or inconsistent use of important terminology hinder progress.

Joern Fischer and David Lindenmayer (2007) recognize two approaches to understanding the effects of landscape modification on species and assemblages that represent the extremes of a continuum of approaches (Figure 21). The first extreme is a species–oriented approach that centres on individual species and assumes that each species responds individualistically to a range of processes related to its requirements for food, shelter, space, and suitable climatic conditions, as well as interspecific processes like competition, predation, and mutualism. An advantage of species-oriented work is its use of well-established ecological causalities, but its key limitation is the impossibility of studying every individual species in any given landscape.

The second extreme is a pattern-oriented approach, where the focus is normally on human-perceived landscape patterns and their correlation with measures of species occurrence, including aggregate measures such as species richness. Pattern-oriented approaches origi-

(*a*) **Pattern-oriented view of a landscape**

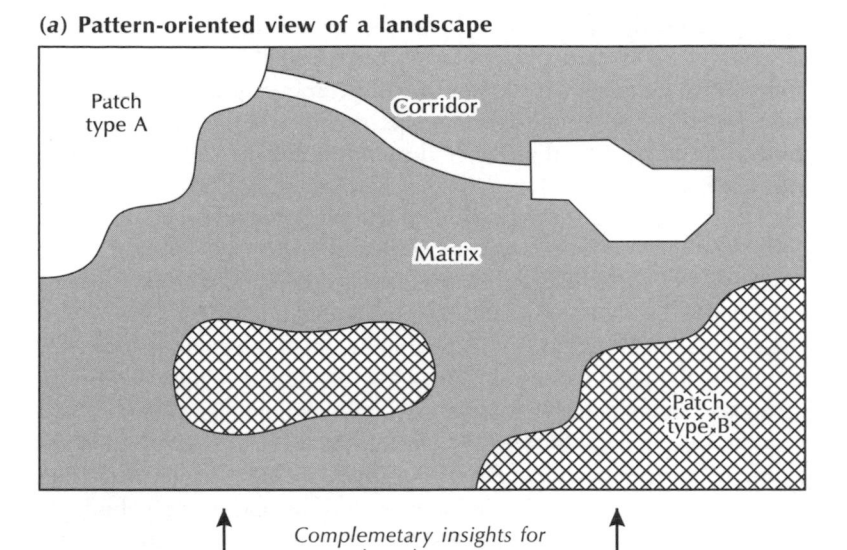

*Complemetary insights for
research and management*

(*b*) **Species-oriented view of a landscape**

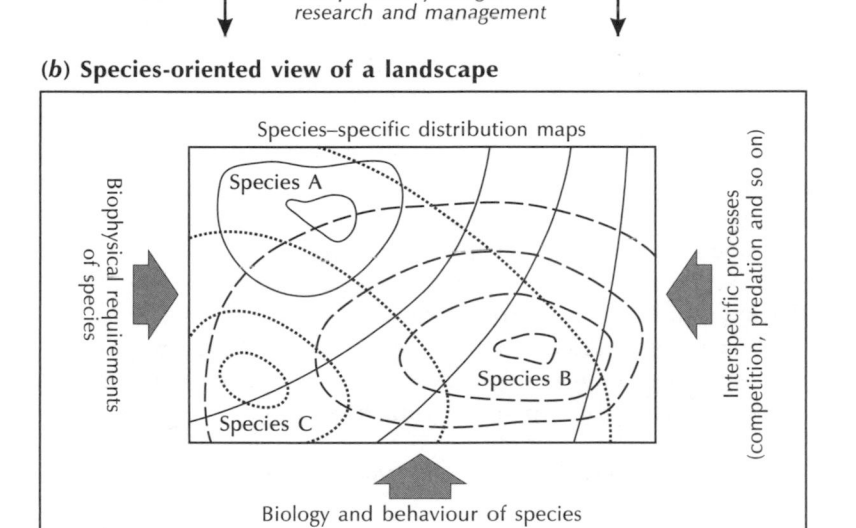

Figure 21 Two chief views of a human-modified landscape.

Source: Adapted from Fischer and Lindenmeyer (2007)

nate from **island biogeography** and are the traditional stronghold of
'fragmentation-related' research (Haila 2002). Two widely used
pattern-oriented conceptual landscape models are the patch–matrix–
corridor model (Wilson and Forman 2008) and to a lesser extent the
variegation model (McIntyre and Barrett 1992; Ingham and Samways

1996) (see **landscape ecology**). All pattern-oriented approaches base themselves on human-defined land cover (usually native vegetation) and seek to establish correlations with species or groups of species to infer potential ecological causalities. They provide broadly applicable general insights, but their chief limitation lies in their practice of aggregating across individual species and ecological processes, which may lead to an oversimplification of the complex casual relationships and subtle differences between individual species.

In spite of their differences, species-oriented and pattern-oriented approaches are complementary in the wider quest of understanding the ecology of modified landscapes, and they have provided some key insights into the processes of habitat loss and habitat fragmentation. In summary, these insights include the following (Fischer and Lindenmayer 2007). **Habitat** is a species-specific concept, and is not restricted to native vegetation. For a given species, threats arise from negative changes to its specific habitat, and disruptions to its biology, behaviour, and interactions with other species. These threats affect vulnerability to **extinction**, which is a common focus of study. A large amount of unmodified native vegetation tends to benefit native species, but there is no universally applicable minimum 'threshold' amount of native vegetation. Edge effects have diverse consequences and variable depth of penetration into patch interiors, but recent reviews suggest they may be less individualistic than previously thought. Conditions in the matrix and landscape heterogeneity are fundamentally important in modified landscapes, and deserve equal attention in research and management as patches of native vegetation. Regime shifts and extinction cascades are particularly likely to arise following a severe reduction in native vegetation cover, a great simplification of vegetation structure throughout the landscape, and the loss of entire functional groups or keystone species.

Further reading: Lindenmayer and Fischer 2006b; Puttker 2008; Steffen *et al.* 2004

HOMEOSTASIS/HOMEORHESIS

These terms come from biology (see Waddington 1957). Homeostasis is a group of system-stabilizing relations characterized by negative **feedback**. Homeorhesis is a group of system-stabilizing relations characterized by positive feedback. Homeostasis involves all those relations that act to preserve a system by keeping it in a steady state during its existence. Homeorhesis involves all those relations that act to preserve, not a steady state, but a flow process that for an individual system follows a relatively fixed trajectory, as is common in growth and development. With homeostasis, system pattern is preserved, so homeostatic change is morphostatic. With homeorhesis, system pattern alters, usually becoming more differentiated and more complex, so homeothetic change is morphodynamic. An example is the set of processes leading to the development of an individual organism, from egg to adult, that operate in conjunction with homeostasis, the processes that maintain an individual in a steady state. That a system may possess both these antithetical facets is something of a paradox; but they are complementary modes of regulation which enable a system to adapt to changes and challenges in its environment.

HYDROLOGICAL CYCLE

The hydrological or water cycle is the circulation of meteoric water (water derived from precipitation) through the hydrosphere, atmosphere, and upper parts of the crust. It links to the circulation of deep-seated, juvenile water associated with magma production and the rock cycle. Juvenile water ascends from deep rock layers through volcanoes, where it issues into the meteoric zone for the first time. On the other hand, meteoric water held in hydrous minerals and pore spaces in sediments, known as connate water, may be removed from the meteoric cycle at subduction sites, where it is carried deep inside the Earth.

The land phase of the water cycle is of special interest to physical geographers. It sees water transferred from the atmosphere to the land and then from the land back to the atmosphere and to the sea. It includes a surface drainage system and a subsurface drainage system. Water flowing within these drainage systems tends to flow within **drainage basins** (watersheds, catchments). The basin water system is, in effect, a set of water stores that receive inputs from the atmosphere and deep inflow from deep groundwater storage, that lose outputs through evaporation and streamflow and deep outflow, and that connect through internal flows. In summary, the basin water runs like this. Precipitation entering the system is stored on the **soil** or rock surface, or is intercepted by vegetation and stored there, or falls directly into a stream channel. From the vegetation it runs down branches and trunks (stemflow), or drips off leaves and branches (leaf and stem drip), or it is evaporated. From the soil or rock surface, it flows over the surface (overland flow), infiltrates the soil or rock, or evaporates. Once in the rock or soil, water may move laterally down hillsides (throughflow, pipeflow, interflow) to feed rivers, it may move downwards to recharge groundwater storage, or it may evaporate. Groundwater may rise by capillary action to top up the rock and soil water stores, or it may flow into a stream (baseflow), or may exchange water with deep storage.

Estimates of annual global precipitation and evaporation fix them both at 973 mm, which converts to 496,100 km³ of water. However, precipitation, evaporation, and runoff are unequal on land and in oceans (Table 6). Annual estimates show that 111,100 km³ of water fall over land, of which 71,400 km³ evaporate and 39,700 km³ is runoff; in the oceans, 385,000 km³ of water fall, 424,700 km³ evaporate, and −39,700 km³ is runoff (this is simply the numerical difference between precipitation and evaporation). This fundamental inequality in the water balance results in a transfer of water vapour (runoff) from the oceans, through the atmosphere, to the continents, and a counter transfer from the continents to the oceans in runoff. These two flows are thus the driving forces behind the water circulation among the latitudinal zones and between continents and oceans. The only regions that do not possess this inequality are continental regions of internal drainage where runoff is zero and precipitation equals evaporation. The fact that the oceanic water exchanges account for 80 per cent of the total highlights the prime importance of the oceans in the global water cycle. Notice that, expressed as water volume, runoff from the land exactly balances the loss from the oceans.

Although but a minute part of the hydrosphere, atmospheric water has a disproportionately large significance to the other environmental **systems**. The volume of water stored in the atmosphere is 13,000 km³. Since the area of the Earth's surface is 510,000,000 km², it is a matter of simple arithmetic to work out that, if all the water vapour in the atmosphere were to condense, it would form a layer 2.54 cm

Table 6 Water balances for land and sea

	Continents (148,900,000 km²)			Oceans (361,100,000 km²)		
	Precipitation	Evaporation	Runoff	Precipitation	Evaporation	Runoff
Water volume (km³)	111,100	71,400	39,700	385,000	424,700	−39,700
Water depth (mm)	746	480	226	1,066	1,176	−110

Source: After Baumgartner and Reichel (1975)

(exactly one inch) deep. Globally, the mean annual precipitation is 97.3 cm. Therefore, there must be 97.3/2.54 = 38 precipitation cycles per year, and the average life of a water molecule in the atmosphere is therefore 365/38 = 10 days. Furthermore, the global store of surface fresh water, if not replenished, would be lost by evaporation in as little as five years and drained by rivers in ten. Table 7 shows the magnitude of the Earth's stores of water and their turnover times.

Table 7 Water stores and their turnover times

Store	Volume (km³)	Turnover time
Biosphere	1,120	Several hours to a week
Atmosphere	12,900	10 days
Rivers	2,120	16 days
Marshes and wetlands	11,500	1–10 years
Lakes and reservoirs	176,400	10–20 years
Soil	16,500	2 weeks to 1 year
Ice caps, ice sheets, and permanent snow	24,064,000	10,000 years
Mountain glaciers	40,600	1,500 years
Oceans and seas	1,338,000,000	2,000–4,000 years
Groundwater	23,400,000	2 weeks to 10,000 years

Source: Adapted from Laycock (1987)

Further reading: Shiklomanov and Rodda 2003

INVASIVE SPECIES

Invasive, exotic, introduced, non-indigenous, alien, and non-native are names given to animal or plant species living outside their normal range. Some researchers dub them 'biological pollution'. Invasions by exotic species of animals and plants are a current and serious environmental problem, and one of the foremost causes of **biodiversity loss**.

The dividing line between native and exotic species can be unclear, even though several criteria indicative of these categories exist. For example, a possible criterion of exotic is 'living outside its natural range owing to human activity'; while a possible criterion of native is 'occurring in its natural range before the arrival of Europeans'. However, complications arise. The first European to visit the Hawaiian Islands was Captain James Cook in 1800, but Polynesians had introduced species to the islands before him. Similarly, some species spread to new areas without the aid of humans: are these native or exotic? Moreover, it is tricky to identify native and exotic species in Europe, Africa, and Asia, where humans have lived for thousands of years but evidence of early introductions is lacking. Such problems as these can be resolved philosophically by accepting that there are several criteria that are typical of native species, and a corresponding set of criteria that are typical of exotic species, but that none of these criteria is either necessary or sufficient to label a species as either native or exotic (Woods and Moriarty 2001).

Some exotic species **populations** may grow to higher levels than in their natural range and cause severe problems for the native animals, plants, and **ecosystems**. They become so abundant because the diseases, parasites, competitors, or predators with which they co-existed in the homeland do not check their numbers. Globally, exotic species have contributed hugely to biodiversity loss over the last few centuries. They are responsible for 42 per cent of reptile **extinctions**, 25 per cent of fish extinctions, 22 per cent of bird extinctions, and 20 per cent of mammal extinctions (Cox 1999). The reasons for this heavy toll are varied. Exotic species may outcompete native species for food and shelter, heavily predate them, adversely alter their **habitats**, precipitate cascade effects in their communities, reduce their genetic integrity by interbreeding with them, and pass on exotic diseases. Exotic competitors sometimes oust native rivals. The exotic rose-ringed parakeet (*Psittacula krameri*) on Mauritius threatens the native Mauritius parakeet (*P. echo*) with extinction because it can exclude it from nest cavities. It is also a potential danger to native species in the United Kingdom, where three main colonies exist in southwest London, southeast London, and the Isle of Thanet. Exotic predators often have profound impacts on native prey species, especially on islands where the native prey species have evolved in the absence of native predators and tend to lack the ability to defend themselves and an innate sense of caution. In particular, exotic rats, cats, and mongooses have played havoc with island native reptile and bird populations. In the Galápagos Islands, native species include endemic reptiles (iguanas and the giant tortoise) and birds, including Darwin's finches. In the 1800s, the first settlers introduced feral cattle, donkeys, horses, goats, pigs, rats, and dogs. Fishermen and visitors later introduced other species. The exotic herbivores overgrazed vegetation, causing the demise of many native plants and prompting the spread of exotic plants. Sometimes, an exotic species alters native habitat in a way that is unfavourable to the native species. Exotic carp have aided the decline of waterfowl in many wetlands, partly through competition, but partly by increasing the turbidity of the water during spawning and feeding, and so lowering the invertebrate populations upon which the native water birds feed. They also destroy aquatic vegetation used by native birds for food, cover, and nesting.

Exotic species may affect native species indirectly by setting in train a cascade of **community changes**. Feral pigs on Hawaii provide an example of this sort of cascade. Once introduced to Hawaii, feral pigs had difficulty in establishing themselves because their diet lacked protein. After the introduction of the earthworm, a ready source of protein became available and the pig population soared. Once abundant, the pigs caused a decline in such native plants as ferns, this decline permitting a rise in the abundance of some exotic plants. A similar kind of cascade effect occurred on the Pacific island of Guam, where the native fauna evolved in the absence of snakes, the island being too remote for reptiles to reach. However, the brown tree snake (*Boiga irregularis*), an efficient nocturnal predator, arrived sometime during World War II, carried in military vehicles and equipment shipped in from New Guinea. Unable to adapt to predation by the brown tree snake, extirpations occurred in thirteen of twenty-two native bird species, two of three native bat species, and four of ten native lizard species.

The interbreeding of exotic and native species reduces the genetic integrity of the native species. In Britain, native red deer (*Cervus elaphus*) and exotic sika deer (*Cervus nippon*) from the Far East tend to interbreed, often producing a hybrid zone where they do so. The fear is that such hybridization may imperil the red deer populations. A botanical example from the British Isles is the native bluebell (*Hyacinthoides non-scripta*) that hybridizes with the introduced Spanish bluebell (*Hyacinthoides hispanica*). Most of the bluebells in London are such hybrids. Introduced species carrying diseases may decimate native populations. In Queensland, Australia, a virus introduced by pet fish seems to have led to a sharp fall in frog populations. In 1998 and 2002, native North Sea seal species – mainly harbour seals and grey seals – suffered population crashes when harp seals (*Phoca groenlandica*) from the Arctic Ocean introduced the distemper virus. Fortunately, the native species have managed to recover.

Finally, not all invasive species are successful. For example, although many of the mammals introduced to New Zealand have thrived, others – bandicoots, kangaroos, racoons, squirrels, bharals, gnus, camels, and zebras – failed to become established.

Further reading: Elton 1958; Mooney *et al.* 2005; Terrill 2007

ISLAND BIOGEOGRAPHY, THEORY OF

The theory of island biogeography combines geographical influences on species diversity with species change, and stresses the dynamism of insular communities. Frank W. Preston (1962), and Robert H. MacArthur and Edward O. Wilson (1963, 1967) proposed it independently. Preston stressed the idea that island species exist in some kind of **equilibrium**. MacArthur and Wilson explicitly set down an equilibrium model. Their central idea was that an equilibrium number of species (animals or plants) on an island is the outcome of a balance between immigration of new species not already on the island (from the nearest area of mainland) and **extinction** of species on the island. In other words, it reflects the interplay of species inputs (from colonization) and species outputs (from extinctions). The equilibrium is dynamic, the consequence of a constant turnover of species. In its simplest form, the MacArthur–Wilson hypothesis makes two key assumptions about what happens to immigration and extinction rates as the number of species living on the island mounts:

1   The rate of species immigration drops (Figure 22a). This happens because, on the average, the more rapidly dispersing species would become established first, causing an initial rapid drop in the overall immigration rate, while the later arrival of slow colonizers would drop the overall rate to an ever-diminishing degree.
2   The rate of extinction of species rises (Figure 22a). It does so because the more species that are present, the more that are likely to go extinct in a unit time.

The point at which the lines for immigration rate and extinction rate cross defines the equilibrium number of species for a given island (Figure 22a).

To refine their model a little, MacArthur and Wilson assumed that immigration rates decrease with increasing distance from source areas: immigration occurs at a higher rate on near islands than on far islands (Figure 22b). For this reason, and all other factors being constant, the equilibrium number of species on a near island will be higher than that on a far island. In addition, MacArthur and Wilson assumed that extinction rates vary inversely with island size: extinction rates on small islands will be greater than extinction rates on large islands (Figure 22b). For this reason, and all other factors being constant, the equilibrium number of species on a small island will be lower than that on a large island. Several other refinements to the model

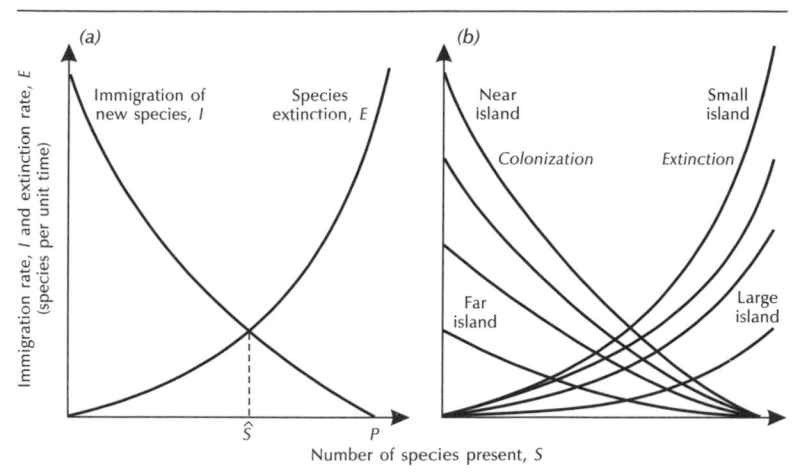

Figure 22 Basic relationships in the MacArthur–Wilson theory of island biogeography. (a) Equilibrium model of a biota on a single island. The equilibrial number of species is defined by the intersection of the curves for the rate of immigration of species not already resident on the island, *I*, and the rate of extinction of species from the island, *E*. (b) Equilibrium models of biotas of several islands lying at various distances from the principal source area of species and of varying size. An increase in distance (near to far) is assumed to lower the immigration rate, while an increase in area (small to large) is assumed to lower the extinction rate.

Source: Adapted from MacArthur and Wilson (1963)

were discussed in the monograph *The Theory of Island Biogeography* (MacArthur and Wilson 1967).

Biogeographers and ecologists raised several objections to the original theory of island biogeography (see Shafer 1990, 15–18). Some claimed that it is so oversimplified as to be useless. Others criticized the failure to include **speciation** on the island, which is probably important on larger islands, the failure to accommodate **habitat** diversity, and the failure to predict the unusual mix of species on islands, which is often very different from the mix of species on neighbouring mainland areas. Despite these criticisms and others, many of which are answerable by minor modifications of the basic model, the theory of island biogeography has engendered much valuable debate and a flood of field investigations into the effects of insularity. Indeed, many studies tend to vindicate the basic thesis that steady-state species numbers are a function of area and distance. For instance, studies of mammalian faunas on islands in Finnish lakes (Hanski 1986) and in the

St Lawrence River (Lomolino 1986) show that they are roughly in equilibrium with colonization rates balancing extinction rates.

The theory of island biogeography has indisputably stimulated new ideas about species in true islands and habitat islands. The latest outgrowths come from Mark Lomolino, who developed a general model of species–area relationships that takes into account the change in factors determining species richness as geographical **scale** increases (Figure 23a) (Lomolino 2000a; Lomolino and Weiser 2001). The idea is that on small islands the dominant factors shaping species richness are such chance events as hurricanes. Beyond an 'ecological **threshold**', on larger islands, such deterministic factors as habitat diversity, **carrying capacity**, and extinction and immigration dynamics as envisioned by MacArthur and Wilson predominate. On large islands lying beyond an 'evolutionary threshold', *in situ* or autochthonous

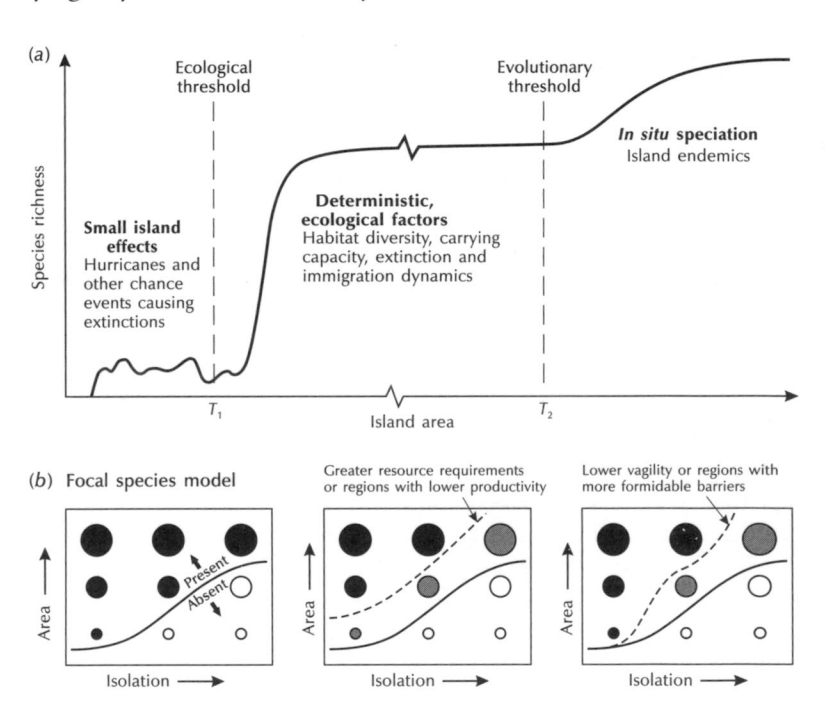

Figure 23 New models of island biogeography. (a) A general model for species–area relationships that includes scale-dependent changes in the chief factors shaping island communities. (b) Insular distribution functions delineating combinations of island area and island isolation where persistence time equals the time between immigration.

Sources: (a) Adapted from Lomolino (2000a); (b) After Lomolino (2000b)

evolution of island endemics is a powerful force bolstering species numbers. Lomolino also built a species-based theory of island biogeography as an alternative to the MacArthur–Wilson theory (Lomolino 2000b). His premise was the idea that general patterns of species assemblages on islands result from, not despite, non-random variation among species. He identified 'insular distribution functions' that separate islands where a species should live from islands where it should not according to island size and island isolation (Figure 23b).

Further reading: Whittaker and Fernadez-Palacios 2006

ISOSTASY

Isostasy is a term used to apply the principle of buoyancy or flotation to continents and oceans. It assumes a state of gravitational **equilibrium** between the lithosphere and underlying asthenosphere such that the tectonic plates 'float' at an elevation dependent on their thickness and density. It helps to explain how different topographic heights can exist at the Earth's surface. When a certain area of lithosphere reaches the state of isostasy, it is said to be in isostatic equilibrium. It is important to note that isostasy is not a process that upsets equilibrium, but rather one that restores it, as when the lithosphere rebounds after ice sheets melt. Some areas (such as the Himalayas) are not in isostatic equilibrium, which has forced researchers to identify other reasons to explain topographic heights (in the case of the Himalayas, by proposing that the force of the impacting Indian plate bolsters their elevation).

Further reading: Watts 2001

LAND DEGRADATION

The degradation of land is a complex problem that affects a significant portion of the Earth's arable areas. Its effects are seen in accelerated soil erosion by wind and water, in soil acidification, alkalization, and salinization, in the loss of soil structure, in the loss of soil organic matter, in derelict soil (soil damaged by industrial or other development and unused). Its effects extend to rivers, wetlands, and lakes that receive the sediments, nutrients, and contaminants from degraded land. The causes of land degradation include land clearance, such as clear-cutting and deforestation, the agricultural depletion of soil nutrients through poor farming practices, livestock, urban conversion, irrigation, pollution, and vehicles travelling off-road. Physical geographers contribute to the study of land degradation and its remediation.

Further reading: Barrow 1991; Bridges *et al.* 2001

LANDSCAPE ECOLOGY

Landscape ecology is hugely successful as a means of exploring spatial aspects of ecological **systems**. It forms a powerful model for viewing the structure and function of the biological and physical landscape that, to some extent, integrates landforms, **soils**, hydrology, and ecology, though it is primarily an ecological construct. It examines how geographical variations in landscapes affect ecological processes, including the distribution and flow of **energy**, materials, and individual organisms, and conversely, how ecological processes affect landscapes. Landscape ecologists have devised their own vocabulary and concepts. Key ideas include **scale**, heterogeneity, patch–corridor–matrix, connectivity, boundary and edge, **ecotones** (ecoclines and ecotopes), **disturbance**, and fragmentation.

A key traditional component of landscape ecology is the patch–corridor–matrix model, which sees a heterogeneous mosaic of **habitat** islands connected by corridors and surrounded by a 'hostile' **environment**. This has had extraordinary success in explaining many features of species patterns and dynamics (e.g. Huggett and Cheesman 2002). Patches, corridors, and matrixes – the landscape elements – are themselves made of individual plants (trees, shrubs, herbs), small buildings, roads, fences, small water bodies, and the like. Moreover, they include natural and human-made landscape components, so the patch–corridor–matrix model integrates the biological and physical aspects of landscapes. Patches are relatively uniform (homogeneous) areas that differ from their surroundings – woods, fields, parks, ponds, rock outcrops, houses, gardens, and so forth. A wealth of quantitative patch metrics describe patches, many of the latest metrics being highly sophisticated and applied to nature conservation problems and nature reserve design (e.g. Bogaert *et al.* 2001a, 2001b). Corridors are strips of land that differ from the land to either side, and link inextricably with patches. They comprise trough corridors (e.g. roads and roadsides, powerlines for electricity transmission, gas lines, oil pipelines, railways, dikes, and trails); wooded strip corridors (e.g. hedgerows and fence-rows); and stream and river (riparian) corridors. Greenways are hybrid corridors comprising parks, trails, waterways, scenic roads, and bike paths. Much research points to a central role played by corridors in landscape ecology and management. For instance, riparian corridors are vital to water and landscape planning and to the restoration of aquatic systems (Naiman and Décamps 1997; Décamps 2001). Matrixes are the background **ecosystems** or land-use types in which patches and corridors are set. The matrix is simply the dominant ecosystem or land-use – forest, grassland, heathland, arable, residential, greenhouses, or whatever – in an area. Its identification is problematic when two or more ecosystems and land-uses co-dominate in a landscape. In these situations, area, connectivity, or other criteria may help to single out the matrix.

The patch–corridor–matrix model is not the only model of landscape ecology. A gradient model may be more appropriate where differences in habitats and species along an environmental gradient make a landscape heterogeneous but not patchy. The variegation model deals with species for which the landscape forms a continuum of migration and that display no qualitative or functional difference between patch or matrix (McIntyre and Barrett 1992).

Above the level of landscape elements are landscape mosaics, within which there is a range of landscape structures. These structures are distinct spatial clusters of ecosystems or land uses or both. Although patches, corridors, and matrixes combine in sundry ways to create landscape mosaics, landscape ecologists recognize six fundamental types of landscape: large-patch landscapes, small-patch landscapes, dendritic landscapes, rectilinear landscapes, chequerboard landscapes, and interdigitated landscapes.

Further reading: Hilty *et al.* 2006; Wiens *et al.* 2006; Wilson and Forman 2008

LIMITING FACTORS AND TOLERANCE RANGE

A limiting factor is an environmental factor that discourages **population** growth. Justus von Liebig (1840), a German agricultural chemist, first suggested the term. He noticed that whichever nutrient happens to be in short supply limits the growth of a field crop. A field of wheat may have ample phosphorus to yield well, but if another nutrient – say nitrogen – should be lacking, then the yield lessens. No matter how much extra phosphorus is applied in fertilizer, the lack of nitrogen will limit wheat yield. Only by making good the nitrogen shortage could yields be improved. These observations led to Liebig to establish a 'law of the minimum': the productivity, growth, and reproduction of organisms will be constrained if one or more environmental factors lies below its limiting level. Later, ecologists established a 'law of the maximum'. This law applies where an environmental factor exceeds an upper limiting level and curtails population growth. In a wheat field, too much phosphorus is as harmful as too little – there is an upper limit to nutrient levels tolerated by plants.

For every environmental factor, such as temperature or moisture, there are three 'zones': a lower limit, below which a species cannot live, an optimum range in which it thrives, and an upper limit, above which it cannot live (Figure 24). The upper and lower bounds define the tolerance range of a species for a particular environmental factor. The bounds vary from species to species. A species will prosper within its optimum range of tolerance; survive but show signs of physiological stress near its tolerance limits; and not survive outside its tolerance range (Shelford 1911).

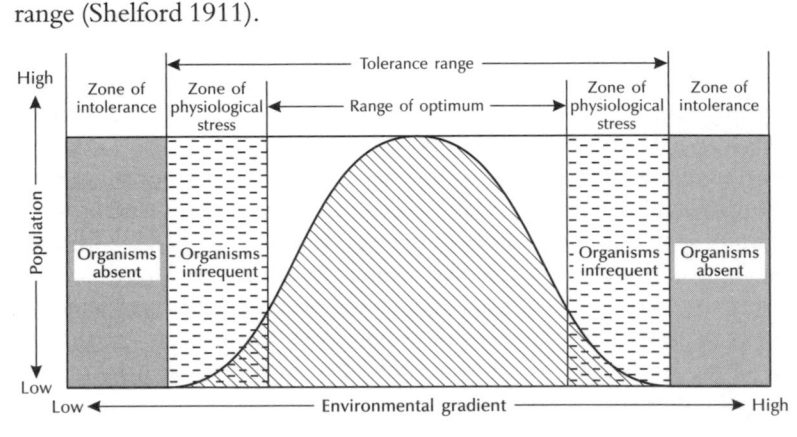

Figure 24 Tolerance range and limits.

Source: Developed from Shelford (1911)

Each species (or race) has a characteristic tolerance range (Figure 25). Stenoecious species have a wide tolerance; euryoecious species have a narrow tolerance. All species, regardless of their tolerance range, may be adapted to the low end (oligotypic), to the middle (mesotypic), or to the high end (polytypic) of an environmental gradient. Take the example of photosynthesis in plants. Plants adapted to cool temperatures (oligotherms) have photosynthetic optima at about 10°C and cease to photosynthesize above 25°C. Temperate-zone plants (mesotherms) have optima between 15°C and 30°C. Tropical plants (polytherms) may have optima as high as 40°C. Interestingly, these optima are not 'hard and fast'. Cold-adapted plants are able to shift their photosynthetic optima towards higher temperatures when grown under warmer conditions.

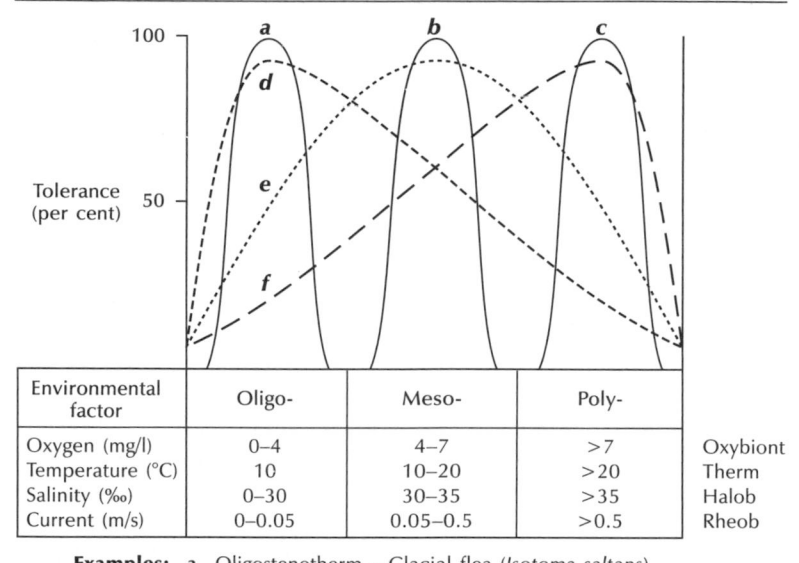

Environmental factor	Oligo-	Meso-	Poly-	
Oxygen (mg/l)	0–4	4–7	>7	Oxybiont
Temperature (°C)	10	10–20	>20	Therm
Salinity (‰)	0–30	30–35	>35	Halob
Current (m/s)	0–0.05	0.05–0.5	>0.5	Rheob

Examples:
a. Oligostenotherm – Glacial flea (*Isotoma saltans*)
b. Mesostenorheob – Barbel (*Barbus fluviatilis*)
c. Polystenoxybiont – Midge (*Liponeura cinerascens*)
d. Oligoeurybiont – Chironomid (*Chironomus plumosus*)
e. Mesoeuryrheob – Freshwater limpet (*Ancylus fluviatilis*)
f. Polyeuryhalob – Jelly-fish (*Aurelia aurita*)

Figure 25 Ecological valency, showing the amplitude and position of the optimum.

Source: Adapted from Illies (1974)

Tolerance may be wide or narrow and the optimum may be at low, middle, or high positions along an environmental gradient. When combined, these contingencies produce six grades of ecological valency (Figure 25). The glacial flea (*Isotoma saltans*), a species of springtail, has a narrow temperature tolerance and likes it cold – it is an oligosteno-therm. The midge *Liponeura cinerascens*, a grazing stream insect, has a narrow oxygen-level tolerance at the high end of the oxygen-level gradient – it is a polystenoxybiont. Figure 25 shows other examples.

LOCAL CLIMATE (TOPOCLIMATE)

Topographic variations over tens of metres to about 100 km produce a mosaic of distinctive local climates (also called topoclimates). Local climates are the climates of the lower troposphere influenced by **topography** (relief, altitude, aspect, water bodies, human features, vegetation, **soils**, and so on). In short, they are the climate of air in contact with the natural and human-made planetary cover.

All topographic features may sufficiently modify radiation fluxes, heat balances, moisture levels, and aerodynamics in the local **environment** to create local climates. Radiation modification depends largely on the aspect and inclination of ground surfaces and the walls and roofs of buildings, on the albedo (the reflectivity of topographic features – different vegetation types, bare soil, human-made surfaces, water bodies), on shading effects, and in some cases, on local **energy** sources (domestic fires, industrial plants, and so on). Some parts of landscapes receive more sunlight than others receive, and emit and absorb different amounts of long-wave radiation. In turn, the altered radiation balances produce hotter and cooler areas within landscapes by modifying local heat balances. Moisture levels vary owing to spatial variability in precipitation receipt (caused by differing interception rates and shelter effects), in evaporation rates, and in soil drainage. Aerodynamic modifications concern the physical effects of small-scale topographic features on airflow.

Urban areas commonly have a distinct local climate. The air in the urban canopy is often warmer than air in the surrounding rural areas, so creating an urban heat island. The precise form and size of urban heat islands are variable and depend upon meteorological, locational, and urban factors. Heat islands occur in tropical and extra-tropical cities. Urban heat-island intensity (the difference between the peak temperature and the background rural temperature) depends on many factors, of which city size is one of the most important. The maximum heat island intensity of Dublin is about 8°C, as is that for Barcelona; it is about 10°C for metropolitan Washington D.C. and 17°C for New York.

Further reading: Akbari 2009; Huggett and Cheesman 2002; Oke 1987

MAGNITUDE AND FREQUENCY

As a rule of thumb, bigger floods, stronger winds, higher waves, and so forth occur less often than their smaller, weaker, and lower counterparts do. Indeed, graphs showing the relationship between the frequency and magnitude of many environmental processes are right-skewed, which means that a lot of low-magnitude events occur in comparison with the smaller number of high-magnitude events, and a very few very high-magnitude events. The return period or recurrence interval expresses the frequency with which an event of a specific magnitude occurs. It is calculable as the average length of time between events of a given magnitude. Take the case of river floods. Observations may produce a dataset comprising the maximum discharge for each year over a period of years. To compute the flood–frequency relationships, the peak discharges are listed according to magnitude, with the highest discharge first. The recurrence interval is then calculated using the equation

$$T = \frac{n+1}{m}$$

where T is the recurrence interval, n is the number of years of record, and m is the magnitude of the flood (with $m = 1$ at the highest recorded discharge). Each flood is then plotted against its recurrence interval on Gumbel graph paper and the points connected to form a frequency curve. If a flood of a particular magnitude has a recurrence interval of 10 years, it would mean that there is a 1-in-10 (10 per cent) chance that a flood of this magnitude (2,435 cumecs – cubic metres per second – in the Wabash River example shown in Figure 26) will occur in any year. It also means that, on average, one such flood will occur every 10 years. The magnitudes of 5-year, 10–year, 25-year,

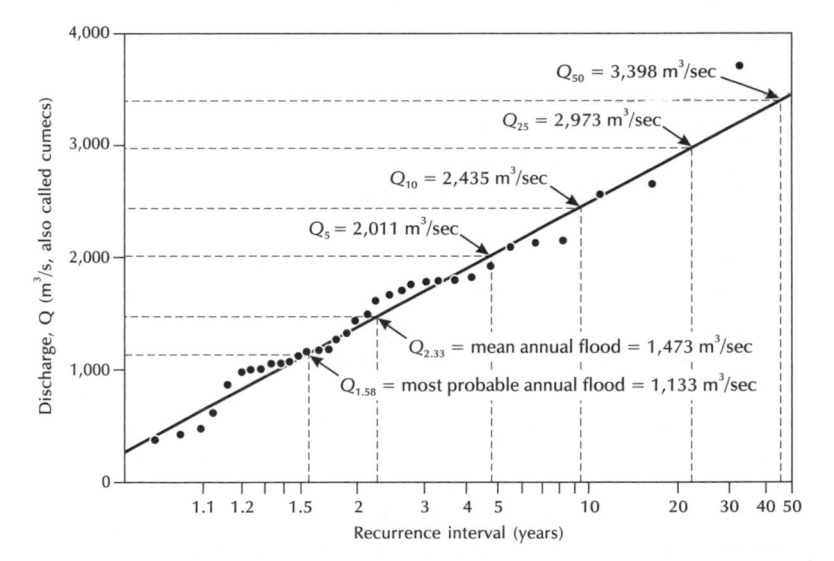

Figure 26 Magnitude–frequency plot of annual floods on the Wabash River, at Lafayette, Indiana, USA. See text for details.

Source: Adapted from Dury (1969)

and 50-year floods are helpful for engineering work, flood control, and flood alleviation. The 2.33-year flood ($Q_{2.33}$) is the mean annual flood (1,473 cumecs in the example), the 2.0-year flood ($Q_{2.0}$) is the median annual flood (not shown), and the 1.58-year flood ($Q_{1.58}$) is the most probable flood (1,133 cumecs in the example).

MASS BALANCE

A mass or material balance applies the law of mass conservation to the analysis of physical **systems**. The mass conservation law dictates that matter cannot be created or destroyed spontaneously; in other words, what goes into a system must be stored, come out, or transform into something else. For mass transactions, the law of mass conservation, sometimes referred to as the continuity condition, leads to the mass storage equation (or continuity of mass equation) which states that

$$\text{change in mass storage} = (\text{mass inputs} - \text{mass outputs}) \times \text{time interval}$$

Physical geographers use mass balances in a range of situations. Hydrologists deal with storage, inputs, and outputs of water, and glaciologists with storage, inputs, and outputs of ice. Geomorphologists focus upon storage, inputs, and outputs of sediments, ecologists upon storage, inputs, and outputs of biochemicals, including pollutants. **Population** ecologists use population 'balances', where change in population numbers ('storage') depends on the number of births and immigrants (population inputs) and on the number of deaths and emigrants (population outputs) over a specified time interval. The theory of **island biogeography** deals with 'species balance' equations.

Useful derivative measures from mass balance work are turnover rate and turnover time. Turnover is the movement of **energy** or mass into, through, and out of a system, and specifically the rate of its depletion and replacement. In a system defined as a set of mass storage compartments, the turnover rate (rate constant, transfer coefficient) is the outgoing flux divided by the steady-state storage. For example, in the Aleut **ecosystem**, some 14.3 per cent of carbon stored in the overlying atmosphere is transferred to surface waters of the ocean; that is a turnover rate of 0.143 (Hett and O'Neill 1974). The turnover time (residence time, time constant, 'life-span') is the reciprocal of the turnover rate. It represents the time that an average 'molecule' stays in the storage compartment. Carbon in the Aleut atmosphere has a turnover time of $1/0.143 = 6.99$ years.

MICROCLIMATE

The microclimate is the climate at or near the ground, in the lowest layer of the atmosphere. Microclimates are also associated with buildings, other human-made surfaces and features, and to such unordinary natural environments as caves and animal burrows. They occupy the vegetation canopy, the urban canopy, the **soil** layer, and buildings. They display horizontal differences over a few centimetres to a hundred metres or so (in clearings). Their vertical dimension is about 1 m in grassland and low crops, to 30 m in forests, and hundreds of metres in some parts of cities and natural landscapes. Microclimates may depend upon several factors. It may be the nature of the substrate (rock, building material, vegetation), the juxtaposition of differing substrates (lakes for instance have a cooling effect on the surrounding land), aspect, and shade effects.

Plants and animals (including humans) use microclimates to their advantage. Well known are the beneficial effects of walls and shelterbelts in agriculture and gardening – they may offer a small region in which crops and plants can grow that would not grow under the regional climate. Less known are the extent to which some animal species utilize microclimatic differences in their **habitats**. A case in point is the daily routine of a colony of yellow-spotted hyrax (*Heterohyrax brucei*) on the Yatta Plateau, southern Kenya, in July 1973 (Vaughan 1978, 431). On first emerging from their nocturnal retreats, deep in rock crevices, these jackrabbit sized mammals avoid touching the cool rock surfaces with their undersides, turn broadside to catch the first rays of the sun, and bask. As the temperature soars during the morning, the hyraxes move to dappled shade beneath the sparse foliage of trees or bushes. When the ambient temperature tops 30°C, they move to deep shade where they lie sprawled on the cool rock and remain there during the hot afternoon. Before dark, they move to the open and lie full length on warm rock.

Further reading: Oke 1987

NATURAL SELECTION

Charles Darwin, in *The Origin of Species* (1859), introduced natural selection, which has since become one of the cornerstones of modern biology. Darwin described natural selection by analogy to artificial selection, a process by which animals with traits desirable to human breeders are systematically favoured for reproduction. He did so in the absence of a valid theory of inheritance, which did not emerge until the early twentieth century. The eventual union of traditional Darwinian **evolution** with later discoveries in classical and molecular genetics created the modern evolutionary synthesis.

Natural selection is a process that affects the fate of heritable traits in successive generations of a **population** of reproducing organisms, leading to favourable heritable traits becoming commoner and to unfavourable becoming rarer. It acts on an organism's phenotype (its observable characteristics – size, shape, colour, and so on). Individuals with favourable phenotypes are more likely to survive and reproduce than those with less favourable phenotypes. If the phenotypes have a genetic basis, then the genotype associated with the favourable pheno-type will increase in frequency in the next generation. Indeed, natural selection is 'the differential perpetuation of genotypes' (Mayr 1970, 107). Over time, this process can result in **adaptation** in which organisms become specialized to fill particular **ecological niches**, and may eventually result in the emergence of new subspecies or even species. Indeed, according to the latest thinking of some biologists, natural selection is a primary driving force of **speciation** and may be more potent than allopatry (the geographical isolation of populations). The argument is that divergent natural selection, which fine-tunes phenotypes to local environments, may outweigh gene flow, leading to further divergence, and so forth, until speciation is accomplished (Dieckmann and Doebeli 1999; Via 2001).

Natural selection tests the genetic foundation of individuals, acting directly on the phenotype and indirectly on the genotype. It may be directional, stabilizing, or disruptive (Figure 27). Directional or progressive selection drives a unidirectional change in the genetic composition of a population, favouring individuals with advantageous characteristics bestowed by a gene or set of genes (Figure 27a). It may occur when a population adapts to a new **environment**, or when the environment changes and a population tracks the changes. The response of the melanic (dark) form of the peppered moth (*Biston betularia*) when confronted with industrial soot illustrates the first case (Kettlewell 1973). The changes in size of the white-tailed deer (*Odocoileus virginianus*) during the Holocene epoch, which appear to track **environmental changes**, exemplify the second type of directional selection (Purdue 1989). Stabilizing selection occurs when a population is well adapted to a stable environment. In this case, selection weeds out the ill-adapted combinations of alleles and fixes those of intermediate character. Stabilizing selection is ubiquitous and probably the most common mode of selection. The peppered moth's

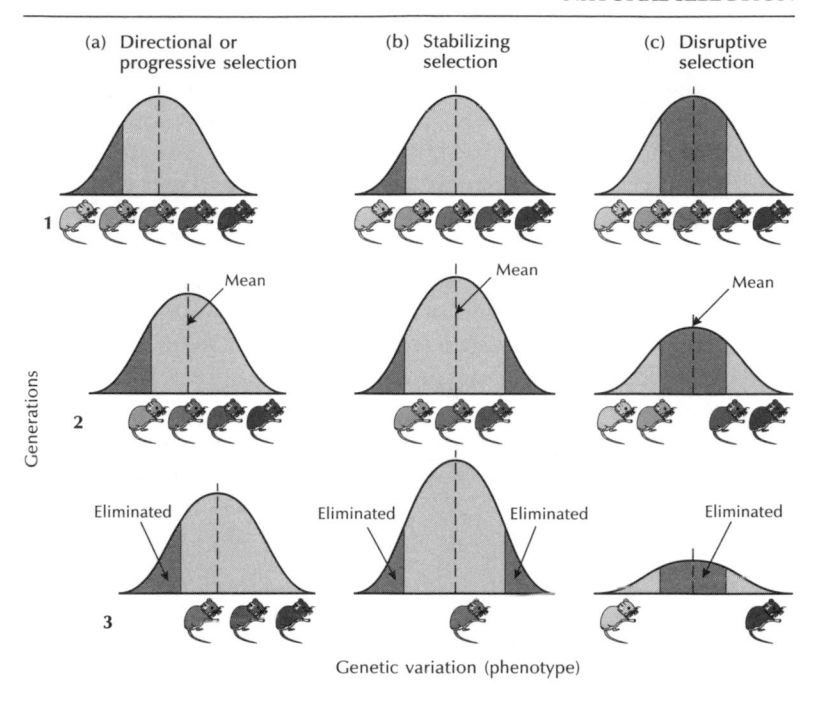

Figure 27 Directional, stabilizing, and disruptive selection.

Source: Adapted from Grant (1977)

response to industrial pollution illustrated directional selection; but the moth population also displays stabilizing selection. Before industrialization altered the moth's environment, stabilizing selection winnowed out the rare melanic mutants, a situation that probably prevailed for centuries. Disruptive or diversifying selection favours the extreme types in a polymorphic population and eliminates the intermediate types, so encouraging polymorphism. At least three situations may promote disruptive selection (Grant 1977, 98–99). The first situation is where well-differentiated polymorphs have a strong selective advantage over poorly differentiated polymorphic types, as in sexually dimorphic species, where males and females that possess distinct secondary sexual characters have a better chance of mating and reproduction than intermediate types (intersexes, homosexuals, and so on). The second situation is where a polymorphic population occupies a heterogeneous **habitat**. The polymorphic types could be specialized for different subniches in the habitat. This may occur in the sulphur butterfly (*Colius euytheme*), the females of which species are polymorphic for wing colour, with one gene controlling orange and white forms. At several localities in California, the white form has an activity peak in the morning and late afternoon, and the orange form has a peak of activity around midday, indicating that the polymorphic types have different temperature and humidity preferences. The third situation occurs when a plant population crosses two different ecological zones. Under these circumstances, different adaptive characteristics may arise in the two halves of the population and persist despite inter-breeding. This appears to be the case for whitebark pine (*Pinus albicaulis*), a high-montane species living at and just above the treeline in the Californian Sierra Nevada (Clausen 1965). On the mountain slopes up to the timberline, the population grows as erect trees; above the timberline, it grows as a low, horizontal, elfinwood form. The arboreal and elfinwood populations are contiguous and cross-pollinated by wind, as witnessed by the presence of some intermediate individuals.

Further reading: Williams 1992

NO-ANALOGUE COMMUNITIES

No-analogue (short for no present or no modern analogue) communities consist of species that exist today, but in combinations not found at present. Other tags for them include disharmonious, mixed, inter-mingled, mosaic, and extraprovincial (Graham 2005). Future communities, including those shaped by current **global warming**, could potentially consist of reshuffled present species and have been called novel or emerging communities (Milton 2003; Hobbs *et al.* 2006). However, they too are no-analogue or disharmonious communities (Williams and Jackson 2007). This usage conflicts with that of Ralf Ohlemüller and his collegues (2006), who use no-analogue to mean current climates with no future analogue, which others call disap-pearing climates (Williams *et al.* 2007).

Past no-analogue communities evolved in, and flourished under, climatic types that no longer exist anywhere in the world. In the southern Great Plains and Texas, USA, present-day grassland or deciduous forest species, such as the least shrew (*Cryptotis parva*), lived cheek-by-jowl with present-day boreal species, such as white-tailed jackrabbit (*Lepus townsendii*) (Lundelius *et al.* 1983). During the late Pleistocene epoch, no-analogue animal communities occurred over all the USA, except for the far west where vertebrate faunas bore a strong resemblance to their modern day equivalents, and date from at least 400,000 years ago to the Holocene epoch. These no-analogue communities evolved from species responding individually to changing environmental conditions during late Pleistocene times (Graham 1979). At the end of the Pleistocene, new environmental changes led to the disassembly of the communities. The climate became more seasonal and individual species had to readjust their distributions. Communities of a distinctly modern mark emerged during the Holocene epoch. In Australia, an Early Pliocene fauna from Victoria – the Hamilton local fauna – contains several extant genera whose living species live almost exclusively in rain forest or rain-forest fringes (Flannery *et al.* 1992). The indication is, therefore, that the Pliocene fauna lived in rain-forest **environment**, but a more complex rain forest than exists today. Taken as a whole, the Hamilton mammalian assemblage suggests a diversity of **habitats** in the Early Pliocene. The environmental mosaic seems to have consisted of patches of rain forest, patches of other wet forests, and open area patches. Nothing like this environment exists today.

PEDOGENESIS

Pedogenesis is the process by which **soil** is created, and is also called soil development, soil **evolution**, and soil formation. Soil formation theory arose in the closing decades of the nineteenth century to become the traditional paradigm of pedology (the scientific study of soils). It originally postulated that soil genesis is the product of downward-acting processes that lead to two sets of interrelated, roughly horizontal, layers – the A horizons and the B horizons, which together constitute the solum. Eluviation washes solutes and fine-grained materials out of the A horizons and deposits them lower down the soil profile in illuvial B horizons. Continued eluviation and illuviation produce coarse-textured residual A and E horizons over heavier-textured B horizons. Under some conditions, organic matter accumulates as distinct O horizons that lie on top of the uppermost A horizon. Later, it was realized that some soil processes mix soil materials and in doing so tend to destroy soil horizons. Pedologists then modified soil formation theory to include the effects of horizon creation (horizonation) and horizon destruction (haploidization) (Hole 1961).

The fathers of pedology – Vasilii Vasielevich Dokuchaev in Russia and Eugene Woldemar Hilgard in the USA – proposed independently that **environmental** factors, especially climate and vegetation, determine the nature of pedogenesis. They did so after noting that soils in different regions but forming under the same climate were the same, as in the case of temperate grassland soils. Hans Jenny (1941) elaborated and refined this view, arguing that nature and rate of pedogenic processes are regulated by 'factors of soils formation' – climate, organisms, relief, parent material, and time. This **functional–factorial approach** to soil genesis, which had a far-reaching impact in many environmental sciences, still is the ruling theory of pedogenesis (see Johnson and Hole 1994).

However, a rival idea – dynamic denudation theory – emerged in the 1990s (Johnson 1993a, 1993b, 2002: Johnson *et al.* 2005; Paton *et al.* 1995). It saw lithospheric material, **topography**, and life (through its role in biomechanical soil processes) as the prime determinative factors of pedogenesis. Its chief tenet was that A horizons are a biomantle (Johnson 1990) created by biomechanical processes, while B horizons are created by epimorphic processes (weathering, leaching, and new mineral formation) acting upon lithospheric material. This means that the textural contrast between A and B horizons is not primarily due to the eluviation and illuviation of fine materials. The role played by biomechanical processes in soil genesis was recognized by Charles Darwin (1881) in his disquisition on earthworms, and by later workers (see Johnson 1993b). Only recently did some researchers rediscover the action of animals and plants as a major factor in soil evolution (e.g. Johnson 1993b; Butler 1995). It is now accepted that the topmost portion of the weathered mantle is subject to mixing by organisms that live in the soil. This mixing is bioturbation. It is mainly caused by the activities of animals (faunal turbation). In soils, earthworms are the most effective bioturbator, followed in order by ants and termites, small burrowing mammals, rodents, and invertebrates. Bioturbation is vigorous enough in almost all environments for the near-surface soil to be described as a bioturbated mantle or biomantle. Some bioturbatory processes produce mounds of bare soil on the ground surface. These little piles of sediment are susceptible of erosion by rain splash and wash processes. The cumulative effect of wash on surface material is a gradual winnowing of fines from the biomantle and a concomitant coarsening of texture.

A new view of pedogenesis emerges by viewing epimorphism and bioturbation together in a three-dimensional landscape (Figure 28). In brief, epimorphism produces saprolite, the upper and finer portion of which is mined by mesofauna to form a topsoil (biomantle). Rainwash further sorts the topsoil and moves it downslope. The result is a soil profile comprising a mobile biomantle, commonly lying on a stone layer, which rests upon subsoil saprolite. The contrast between topsoil and subsoil is often seen in soil texture – the biomantle is dominated by residual quartz and displays few features of the bedrock.

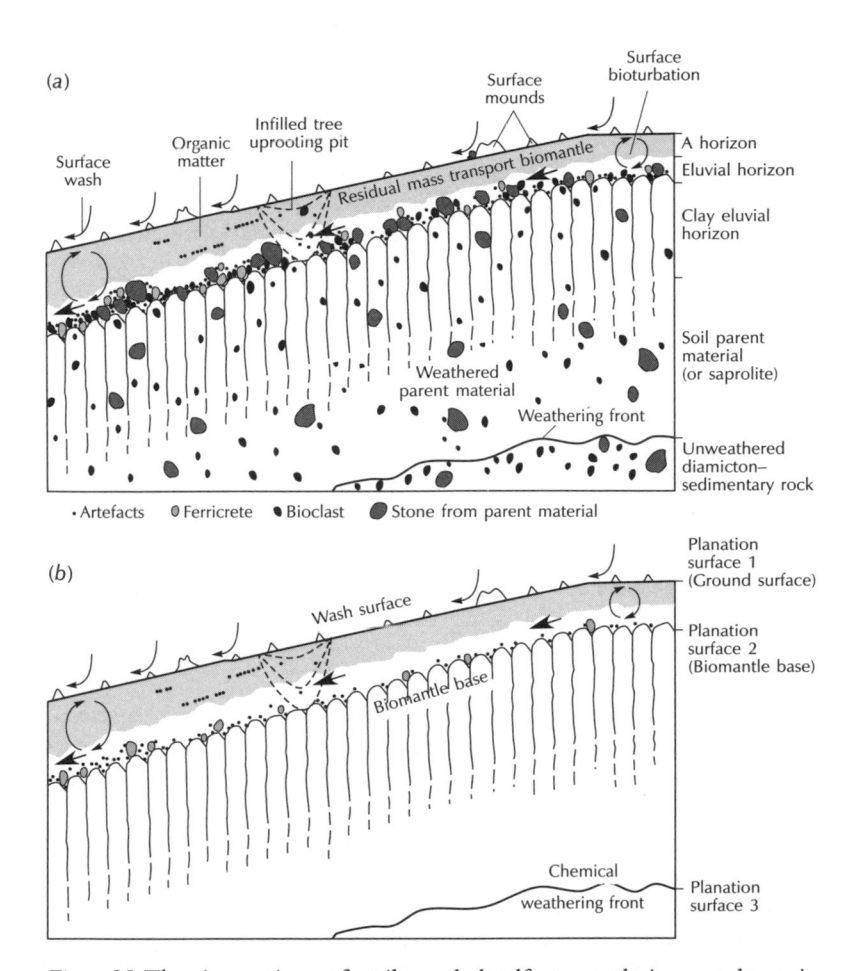

Figure 28 The interaction of soils and landform evolution – dynamic denudation. (a) Dynamic denudation in landscapes underlain by stony sediments. (b) Dynamic denudation in landscapes underlain by stone-free sedimentary rocks. In both cases, three planation surfaces are present. The first is the chemical weathering or dissolution front that migrates downward with time. Material released by dissolution is carried away laterally by groundwater and throughflow. The second is the dominant wash surface (the ground surface). Disturbance by animals and plants is great. Rain and wind carry fine materials downslope, the finest particles moving the farthest. The third is the boundary between the topsoil (A and E horizons) and the subsoil (B horizons). It separates the biomantle from largely *in situ* parent material and is commonly marked by a stone-line or metal nodules. Much of the soluble material washed out of the biomantle is carried downslope by throughflow just above this boundary.

Source: Adapted from Johnson (1993b)

Rarely, the contrast is seen in fabric. Aeolian processes add to the development of the texture contrast by winnowing fines and leaving behind coarser, well-sorted topsoils. The residue may be capable of forming mobile sand dunes. Thus the wind is responsible for the accumulation of quartz as the final residual product of pedogenesis. This is evident in Africa and Australia at present. It was far more common a process in the past.

Further reading: Schaetzl and Anderson 2005

PLATE TECTONICS

The plate tectonic model is the ruling theory for explaining changes in the Earth's crust. This model is thought satisfactorily to explain geological structures, the distribution and variation of igneous and metamorphic activity, and sedimentary facies. In fact, it seems to explain all major aspects of the Earth's long-term tectonic **evolution**. The plate tectonic model comprises two tectonic 'styles'. The first involves the oceanic plates and the second involves the continental plates.

The oceanic plates are part of the cooling and recycling system comprising the mesosphere, asthenosphere, and lithosphere beneath the ocean floors (Figure 29). The chief cooling mechanism is subduction. New oceanic lithosphere forms through volcanic eruptions along mid-ocean ridges. The newly formed material moves away from the ridges. In doing so, it cools, contracts, and thickens. Eventually, the oceanic lithosphere becomes denser than the underlying mantle and sinks. The sinking takes place along subduction zones, which are associated with earthquakes and volcanicity. Cold oceanic slabs may sink well into the mesosphere, perhaps as much as 670 km or below the surface. Indeed, subducted material may accumulate to form 'lithospheric graveyards' (Engebretson *et al.* 1992). It is uncertain why plates should move. Several driving mechanisms are plausible. Basaltic lava upwelling at a mid-ocean ridge may push adjacent lithospheric plates to either side. On the other hand, as elevation tends to decrease and slab thickness to increase away from construction sites, the plate may move by gravity sliding. Another possibility, currently thought to be the primary driving mechanism, is that the cold, sinking slab at subduction sites pulls the rest of the plate behind it. In this scenario, mid-ocean ridges stem from passive spreading – the oceanic lithosphere is stretched and thinned by the tectonic pull of older and denser lithosphere sinking into the mantle at a subduction site; this would explain why the sea floor tends to spread more rapidly in plates attached to long subduction zones. As well as these three mechanisms, or perhaps instead of them, mantle **convection** may be the number one motive force, though this now seems unlikely, as many spreading sites do not sit over upwelling mantle convection cells. If the mantle-convection model were correct, mid-ocean ridges should display a consistent pattern of gravity anomalies, which they do not, and would probably not develop fractures (transform faults). But, although convection is perhaps not the master driver of plate motions, it does occur. There is some disagreement about the depth of the convective cell – is it confined to the asthenosphere, the upper mantle, or the entire mantle (upper and lower)? Whole mantle convection (see Davies 1999) has gained much support, although it now seems that

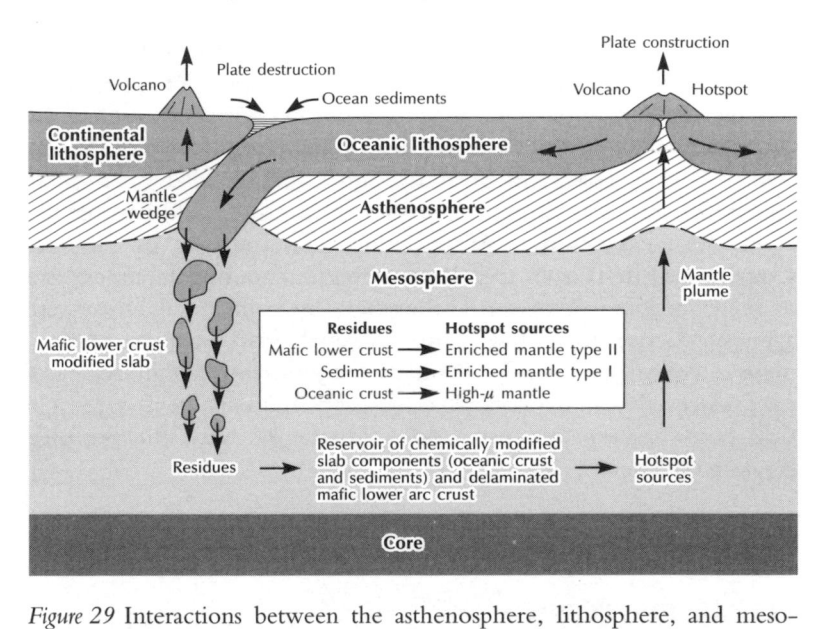

Figure 29 Interactions between the asthenosphere, lithosphere, and meso-sphere. The oceanic lithosphere gains material from the mesosphere (via the asthenosphere) at constructive plate boundaries and hotspots and loses material to the mesosphere at destructive plate boundaries. Subduction feeds slab material (oceanic sediments derived from the denudation of continents and oceanic crust), mantle lithosphere, and mantle wedge materials to the deep mantle. These materials undergo chemical alteration and accumulate in the deep mantle until mantle plumes bear them to the surface where they form new oceanic lithosphere.

Source: Adapted from Tatsumi (2005)

whole mantle convection and a shallower circulation may both operate.

The continental lithosphere does not take part in the mantle convection process. It is 150 km thick and consists of buoyant low-density crust (the tectosphere) and relatively buoyant upper mantle. It therefore floats on the underlying asthenosphere. Continents break up and reassemble, but they remain floating at the surface. They move in response to lateral mantle movements, gliding serenely over the Earth's surface. In breaking up, small fragments of continent some-times shear off; these are called terranes. They drift around until they meet another continent, to which they become attached (rather than being subducted) or possibly are sheared along it. As they may come from a different continent than the one to which they abut, they are called exotic or suspect terranes. Most of the western seaboard of North America appears to consist of these exotic terranes. In moving, continents have a tendency to drift away from mantle hot zones, some of which they may have produced: stationary continents insulate the underlying mantle, causing it to warm. This warming may eventually lead to a large continent breaking into several smaller ones. Most continents are now sitting on, or moving towards, cold parts of the mantle. An exception is Africa, which was the core of Pangaea. **Continental drift** leads to collisions between continental blocks and to the overriding of oceanic lithosphere by continental lithosphere along subduction zones. Geologists have established the relative move-ment of continents over the Phanerozoic aeon with a high degree of confidence, although pre-Pangaean reconstructions are less reliable than post-Pangaean reconstructions. Figure 30 charts the probable breakup of Pangaea.

Further reading: Johnson 2006; Oreskes 2003

PLUME TECTONICS

W. Jason Morgan (1971) proposed mantle plumes as geological features. In doing so, he extended J. Tuzo Wilson's (1963) idea of hotspots, which Wilson used to explain the time-progressive forma-tion of the Hawaiian island and seamount train as the Pacific sea floor moved over the Hawaiian hotspot lying atop a 'pipe' rooted to the deep mantle. Mantle plumes may be hundreds of kilometres in diam-eter and rise towards the Earth's surface from the core–mantle boundary or from the boundary between the upper and lower mantle. A plume consists of a leading 'glob' of hot material followed by a

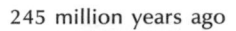
245 million years ago

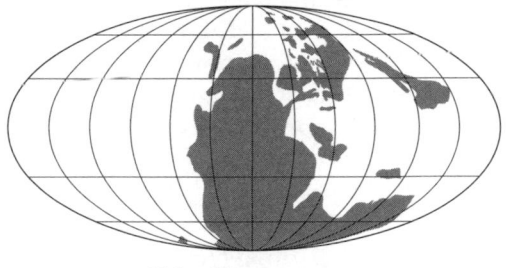

160 million years ago

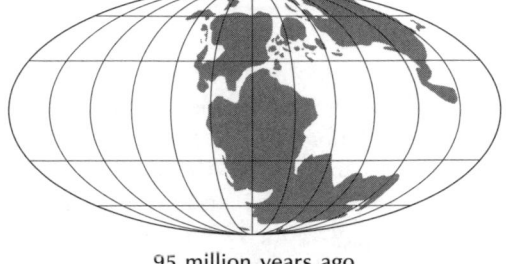

95 million years ago

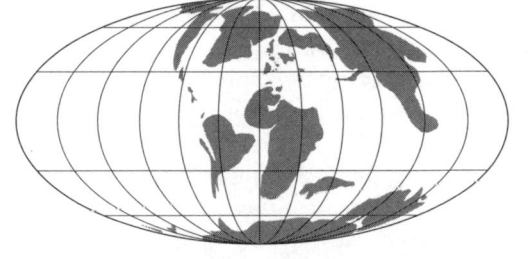

30 million years ago

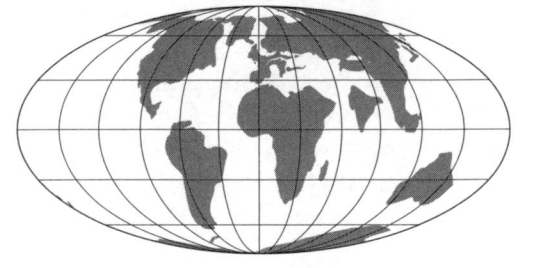

Figure 30 Changing arrangement of continents over the last 245 million years,
showing the breakup of Pangaea, during the Early Triassic period;
during the Callovian age (Middle Jurassic); during the Cenomanian
age (Late Cretaceous); and during the Oligocene epoch. All maps
use Mollweide's equal-area projection.

Source: Adapted from maps in Smith *et al.* (1994)

'stalk'. On approaching the lithosphere, the plume head mushrooms beneath the lithosphere, spreading sideways and downwards a little. The plume temperature is 250–300°C hotter than the surrounding upper mantle, so that 10–20 per cent of the surrounding rock melts. This melted rock may then run onto the Earth's surface as flood basalt. The mechanisms by which mantle plumes form and grow are undecided. They may involve rising plumes of liquid metal and light elements pumping latent heat outwards from the inner-core boundary by compositional **convection**, the outer core then supplying heat to the core–mantle boundary, whence giant silicate magma chambers pump it into the mantle, so providing a plume source (Morse 2000).

Researchers disagree about the number of plumes, typical figures being 20 in the mid-1970s, 5,200 in 1999 (though these include small plumes that feed seamounts), and nine in 2003 (Foulger 2005). Plumes come in a range of sizes, the biggest being megaplumes or super-plumes. A superplume may have lain beneath the Pacific Ocean during the middle of the Cretaceous period (Larson 1991). It rose rapidly from the core–mantle boundary about 125 million years ago. Production tailed off by 80 million years ago, but it did not stop until 50 million years later. It is possible that cold, subducted oceanic crust on both edges of a tectonic plate accumulating at the top of the lower mantle causes superplumes to form. These two cold pools of rock then sink to the hot layer just above the core, squeezing out a giant plume between them (Penvenne 1995).

Some researchers speculate that plume tectonics may be the dominant style of convection in the major part of the mantle. Two super-upwellings (the South Pacific and African superplumes) and one super-downwelling (the Asian cold plume) appear to prevail (Figure 31), which influence, but are also influenced by, **plate tectonics**. Indeed, crust, mantle, and core processes may act in concert to create 'whole Earth tectonics' (Maruyama 1994; Maruyama et al. 1994; Kumazawa and Maruyama 1994). Whole Earth tectonics integrates plate tectonic processes in the lithosphere and upper mantle, plume tectonics in the lower mantle, and growth tectonics in the core, where the inner core slowly grows at the expense of the outer core. Plate tectonics supplies cold materials for plume tectonics. Sinking slabs of stagnant lithospheric material drop through the lower mantle. In sinking, they create super-upwellings that influence plate tectonics, and they modify convection pattern in the outer core, which in turn determines the growth of the inner core.

Despite the alluring plausibility of the plume tectonic model, a minority of rebellious voices have always spoken out against plumes, and, since about the turn of the millennium, the number of voices has swollen and the validity of the plume model has emerged as a key debate in Earth science (see Anderson 2005; Foulger 2005).

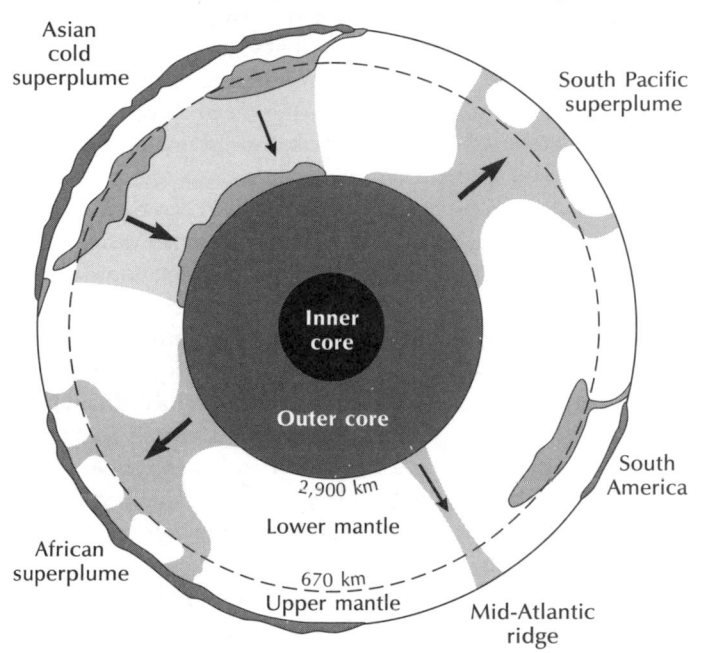

Figure 31 A possible grand circulation of Earth materials. Oceanic lithosphere, created at mid-ocean ridges, is subducted into the deeper mantle. It stagnates at around 670 km and accumulates for 100–400 million years. Eventually, gravitational collapse forms a cold downwelling onto the outer core, as in the Asian cold superplume, which leads to mantle upwelling occurring elsewhere, as in the South Pacific and African hot superplumes.

Source: Adapted from Fukao *et al.* (1994)

Further reading: Foulger *et al.* 2005; Yuen *et al.* 2007

POPULATIONS/METAPOPULATIONS

The idea of a population is central to much work in biology and ecology. A population is a loose collection of individuals of the same species. Red deer (*Cervus elaphas*) in Britain constitute a population. All of them could interbreed, should the opportunity arise. In practice, most populations, including the red deer population in Britain, exist as sets of local populations or demes. A local red deer population lives in the grounds of Lyme Park, Cheshire, its members forming a tightly linked, interbreeding group. Population biologists have developed sophisticated models that, given a starting population age-structure and birth and survival rates for each age and sex class in the population, enable the prediction of population changes. Such models have proved salutary in studying conservation and management issues, for example in assessing the best way of conserving the endangered piping plover (*Charadrius melodus melodus*) population in eastern Canada (Calvert *et al.* 2006).

Useful though classic population models may be, ecologists found it helpful to consider the distribution of populations in landscapes and, in 1969, Richard Levins came up with the idea of metapopulations. These are basically subpopulations of the same species living in different parts of a landscape and interacting with each other to varying degrees. The degree and nature of the interaction varies and this has led to the recognition of four chief kinds of metapopulation, which are normally called 'broadly defined' (or true), 'narrowly defined', 'extinction and colonization', and 'mainland and island' (Gutiérrez and Harrison 1996), although the broad and narrow categories might be more concisely named 'loose' and 'tight' metapopulations, respectively (Figure 32).

A loose metapopulation is simply a set of subpopulations of the same species, all of which are susceptible of extirpation (local **extinction**). Rates of mating, competition, and other interactions are much higher within the subpopulations than they are between the subpopulations. The subpopulations may arise through an accident of geography – the **habitat** patches in which the subpopulations live lie farther apart than the normal **dispersal** distance of the species – and they may or may not be interconnected. By this definition, most species with large and discontinuous distributions form metapopulations. Where habitat patches are so close together that most individuals visit many patches in their lifetime, the subpopulations behave as a single population – all individuals effectively live together and interact. Where habitat patches are so scattered that dispersal between them almost never happens, the subpopulations behave effectively as separate populations. This situation applies to mammal populations living on desert mountaintops in the American Southwest (Brown 1971).

A tight or narrowly defined metapopulation is a set of subpopulations of the same species living in a mosaic of habitat patches, with a significant exchange of individuals between the patches. Migration or dispersal among the subpopulations stabilizes local population fluctuations. In theory, a tight metapopulation structure occurs where the

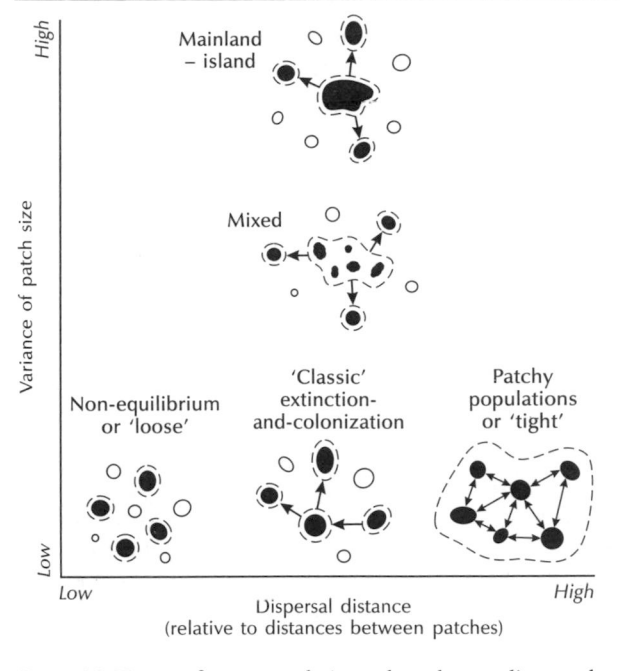

Figure 32 Types of metapopulations classed according to the variance of patch size and the dispersal distance.

Source: After Huggett (2004)

distance between habitat patches is shorter than the species is physically capable of travelling, but longer than the distance most individuals move within their lifetimes. An example is the European nuthatch (*Sitta europaea*) population. The nuthatch inhabits mature deciduous and mixed forest and has a fragmented distribution in the agricultural landscapes of Western Europe (Verboom *et al.* 1991). It shows all the characteristics of a tight metapopulation: the distribution is dynamic in space and time; the extinction rate depends on patch size and habitat quality; and the patch colonization rate depends on the density of surrounding patches occupied by nuthatches.

In the narrowest possible sense, and as originally defined (Levins 1970), a metapopulation consists of subpopulations characterized by frequent turnover. In these 'extinction-and-colonization' metapopulations, all subpopulations are equally susceptible to local extinction. Species persistence therefore depends on there being enough subpopulations, habitat patches, and dispersal to guarantee an adequate rate of recolonization. This metapopulation structure seems to apply to the pool frog (*Rana lessonae*) on the Baltic coast of east-central Sweden (Sjögren 1991). Frog immigration mitigates against large population fluctuations and inbreeding, and 'rescues' local populations from extinction. Extinction-and-colonization metapopulation structure also applies to an endangered butterfly, the Glanville fritillary (*Melitaea cinxia*) in a fragmented landscape (Hanski *et al*. 1995).

In some cases, there may be a mixture of small subpopulations prone to extinction and a large persistent population. The viability of the 'mainland–island' or 'core–satellite' metapopulations is less sensitive to landscape structure than in other types of metapopulation because it has at least one subpopulation immune to extirpation. An example is the natterjack toad (*Bufo calamita*) metapopulation at four neighbouring breeding sites in the North Rhineland, Germany (Sinsch 1992) and the bay checkerspot butterfly (*Euphydryas editha bayensis*) in California, USA (Harrison *et al*. 1988).

Metapopulation theory is now widely applied in conservation biology. It is especially important where **habitat fragmentation** is the central concern. A 'classic' example is the northern spotted owl in the United States, for which bird the metapopulation approach to management has proved beneficial (e.g. Noon and Franklin 2002).

Further reading: Hanski 1999; Neal 2004

REFUGIA

In biogeography, refugia (singular: refugium) are places with isolated or relict **populations** of animal or plant species that were formerly more widespread. Climatic changes or such human activities as deforestation and overhunting may have caused the isolation. Present examples of refuge species are the mountain gorilla, isolated to specific mountains in central Africa, and the Australian sea lion, isolated to specific breeding beaches in South Australia due to overhunting. In many cases, the isolation of refugia is transitory, but some refugia may endure long enough to house many endemic species surviving as relict populations.

The notion of refugia has aided the understanding of the fate of temperate fauna and flora during the glacial climates that have characterized some four-fifths of the last two million years of Earth history. An argument holds that they survived in refugia where they could endure the adverse conditions. The exact location of these more favourable pockets is still under investigation (e.g. Willis and

Whittaker 2000). One study using macroscopic charcoal, in conjunction with analyses of pollen and molluscs, showed that at least seven tree species grew in Hungary between about 32,500 and 16,500 radiocarbon years BP (Willis *et al.* 2000). They probably eked out a living in favourable microenvironments that provided an important cold-stage refugium for the European fauna and flora. Indeed, such refugia for temperate species appear to have exerted a strong influence over current patterns of **biodiversity** in that continent. For instance, the distinctive patterns of genetic variation in present-day populations of grasshoppers, hedgehogs, the brown bear, house mice, voles, oaks, common beech, black alder, and silver fir seemed to be linked to isolation in cold-stage refugia in Spain, southern Italy, and the Balkans (Willis and Whittaker 2000). In short, the cold-stage isolation of these species has left a mark on present-day biodiversity (see also **vicariance**).

In the tropics, refugia in the Amazon Basin – isolated pockets of tropical forest surrounded by lowland savannah – may have acted as 'species pumps' that generated new species, which model seems to account for the high numbers of endemic species in many tropical regions. Jürgen Haffer (1969) first proposed this idea to explain the high biodiversity of bird populations in the Amazonian river basin. He suggested that a change to more arid conditions in the late Pleistocene produced isolated patches of tropical forests in which bird populations survived. Over time, allopatric **speciation** occurred in the patches, creating sister species. At the end of the Pleistocene, more humid conditions returned, the forest expanded, and the refugia were reconnected.

Curiously, while the refugia hypothesis is gaining acceptance as an explanation of temperate latitude diversity patterns, it is going out of favour as an explanation of tropical diversity. Pollen evidence suggests that lowland tropical forests were not widely replaced by savannah during glacial stages, and that forests persisted over much of the lowland Amazon Basin (Colinvaux *et al.* 1996). Sea-level rises of around 100 m during interglacial stages are a possible cause of the patterns of endemic species in the Amazon (Nores 1999). Sea-level rises of that magnitude would have fragmented the region into two large islands and several smaller archipelagos. The insular populations would have been isolated so promoting allopatric speciation during warm stages.

Further reading: Weiss and Ferrand 2007

REGION

In physical geography and its cognate disciplines, a region is normally a spatial unit of some feature – community, **ecosystem**, slope, **drainage basin**, **soil**, sea, or whatever. It is characteristically smaller than an entire area of interest (such as global vegetation and global soils), but larger than its component units. Its definition rests upon several features rather than a single feature, and produces uniform (homogeneous or formal) regions or functional (nodal) regions. In ecology, a nested hierarchy of formal units covers a range of spatial scales. Small units go by various names, including sites, micro–ecosystems, land types, and land units. These small units form landscape mosaics, meso–ecosystems, land-type associations, subregions, and so on. In turn, these medium-sized spatial units form larger units, variously styled regions, **ecoregions**, provinces, divisions, domains, zones, ecozones, kingdoms, and so forth. These units are definable as functional regions, where flows help to sustain the integrity of the unit.

Take as an example regional units in hydrology. There are many systems for defining functional surface-water regions. The drainage basin is the fundamental stream-based region. It forms the basis of hierarchical systems of hydrological units. In the USA, the 'Federal Standard for Delineation of Hydrologic Unit Boundaries' delineates hydrologic units in a six-level hierarchy covering the entire country. Starting with the largest unit, these are regions, subregions, basins, sub-basins, watersheds, and subwatersheds. In the USA, there are 21 hydrologic unit regions, 222 subregions, 352 basins, and 2,149 sub-basins, and an estimated 22,000 watersheds and 160,000 subwatersheds.

RESILIENCE

In general terms, resilience is the ability of a system to recover from, or to resist being affected by, a **disturbance**. Resilience is definable in at least two ways. First, in the engineering definition, it is the rate at which a system returns a single steady or cyclical state following a perturbation. This definition assumes that the system always stays within the stable domain that contains the steady state (attractor or **equilibrium** point). Second, in the ecological definition applied to **systems** that have more than one stability domain (several attractors, multiple equilibria), resilience is the amount of disturbance needed to flip the system from one stability domain to another. By this definition, resilience measures the ability of an **ecosystem** to persist in the face of perturbations arising from weather events, fire, pollution, **invasive species**, humans, and so forth; or, put another way, how much disturbance an ecosystem can absorb before it flips to a new configuration (Holling 1973). Coral reefs display ecological resilience in the face of frequent disturbance (Woodroffe 2007), and many reefs seem to be a temporal mosaic of communities at different stages of recovery from these assorted short-term disturbances (McManus and Polsenberg 2004). Ecosystems may be highly resilient and yet have low stability. For instance, bird and mammal **populations** in the boreal forest community of central Canada and Alaska fluctuate greatly in line with the ten-year cycle of snowshoe hare numbers, but the community has lasted for centuries.

In geomorphology, resilience is the degree to which a system recovers to its original state following a disturbance (Brunsden 2001). This definition accommodates the fact that some landform systems may not return to an identical state after a disturbance owing to some kind of barrier (possibly a transient one) impeding the change between alternative states. It recognizes that landform systems are seldom stationary in time, as boundary conditions change at various timescales. For example, in coastal systems, sea level changes over several timescales and may have a profound impact on coastal landforms.

SCALE

Environmental change occurs across all spatial scales, from a cubic centimetre to the entire globe, and across all timescales, from seconds to aeons. Space and time are continuous. However, it is normal practice to slice Earth's spatial and temporal dimensions into convenient portions and to label them. No agreed labelling system exists. The prefixes micro, meso, macro, and mega are very popular. They produce the expressions microscale, mesoscale, macroscale, and megascale. It is far simpler, if far less impressive, to use the terms small-scale, medium-scale, large-scale, and very large-scale. The boundaries between the categories are somewhat arbitrary. How large does small have to be before it becomes medium? A consensus of sorts has emerged, though the exact divisions adopted depend to some extent upon what it is that changes – climate, **soils**, slopes, landscapes, and so forth. It is possible that the **environment** has a fractal structure. If this were the case, there would be natural divisions of time and space scales. This question is under investigation. Until a sure conclusion emerges, it seems best to create divisions using logical steps, say whole powers of ten. Figure 33 shows a possible scheme, in which the spatial

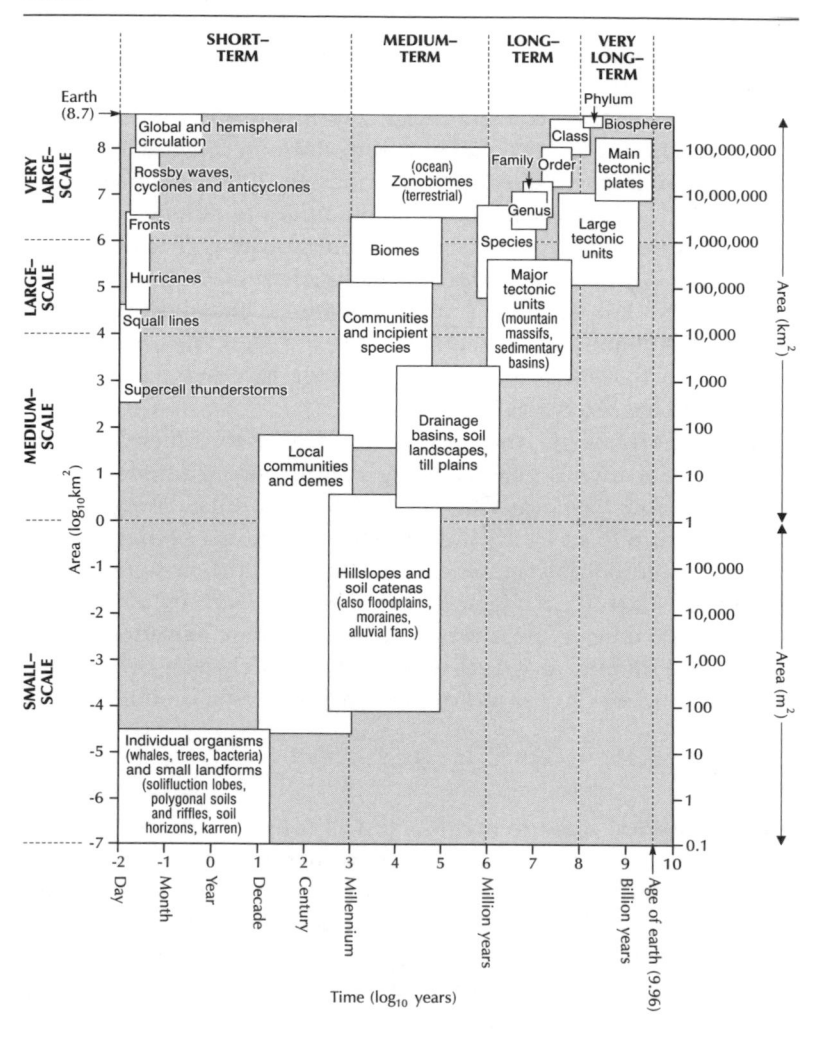

Figure 33 Scales of environmental change.

Source: Adapted from Huggett (1991, 1997a)

dimension has the following limits: microscale is up to 1 km²; meso-scale is 1–10,000 km²; macroscale is 10,000–1,000,000 km²; and megascale is from 1,000,000 km² to the entire surface area of the Earth.

The micro–meso–macro–mega designation is applicable to the time dimension. However, the terms short-term, medium-term, long-term, and very long-term seem plainer. As with spatial divisions, so with temporal ones: it is a matter of personal preference where to set the limits between the time domains. A possible division is: short-term up to 1,000 years, medium-term from 1,000 to 1,000,000 years, long-term from 1,000,000 to 100,000,000 years, and very long-term from 100,000,000 to 4,600,000,000 years. The terms 'megayear' (= 1 million years) and 'gigayear' (= 1 billion (US) = 1,000 million years) are useful when talking of long-term and very long-term environmental changes.

Components of the environment relate to these spatial and temporal scales. Each component has a size and a 'life-span'. Figure 33 shows some of the components. Notice that atmospheric **systems**, be they local or global, survive for less than a year. This is due to the (literally) fluid nature of air. Tectonic units of the lithosphere may survive for billions of years. Soils, landscapes, and communities are intermediate between atmosphere and lithosphere, surviving for anything from centuries to hundreds of megayears.

SEA-LEVEL CHANGE

Sea level seldom remains unchanged for long owing to volumetric and mass distribution changes in the oceans. Ocean volume changes are eustatic or steric. Eustatic change results from water additions or extractions from the oceans (glacio-eustatic change) and from changes in ocean-basin volume (tectono-eustatic change). **Tectonics** is the ultimate control of sea level: in the case of tectono-**eustasy**, the control is direct; in the case of glacio-eustasy, the control is indirect because tectonics (and other factors) alter climate and climate in turn alters sea level. Steric change results from temperature or density changes in seawater. Much of the predicted sea-level rise during the current century will result from the thermal expansion of seawater, which is about 20 cm/°C per 1,000 m.

Sea level fluctuates over all timescales. Medium-term and long-term changes are recorded in sedimentary rocks and revealed by the technique of seismic stratigraphy (see Coe *et al.* 2003). They display six superimposed orders of cyclical sea-level change during the Phanerozoic aeon, each cycle having a distinct signature. The changes in sea level resulting from the first- and second-order cycles can be as much as 250 m. Such very long-term changes of sea level provide a useful yardstick against which to discuss Quaternary sea-level changes and predicted rises of sea level over the current century.

During the Quaternary, sea level at times stood higher and at other times lower than at present. These highstands and lowstands of sea level have left evidence in landscapes, with marine terraces recording highstands and drowned landscapes recording lowstands. Glacio-eustatic mechanisms were the chief drivers of the changes as glacial conditions, when water was trapped in ice, alternated with interglacial conditions. In detail, various types of raised shoreline – stranded beach deposits, beds of marine shells, ancient coral reefs, and platforms backed by steep, cliff-like slopes – all attest to higher stands of sea level. Classic examples come from fringing coasts of formerly glaciated areas, such as Scotland, Scandinavia, and North America. An example is the *Patella* raised beach on the Gower Peninsula, South Wales (Bowen 1973). A shingle deposit lies underneath tills and periglacial deposits associated with the last glacial advance. The shingle is well cemented and sits upon a rock platform standing 3–5 m above the present beach. It probably formed around 125,000 years ago during the last inter-glacial stage, when the sea was 5 m higher than now. Ancient coral reefs sitting above modern sea level are also indicative of higher sea levels in the past. In Eniwetok atoll, the Florida Keys, and the Bahamas, a suite of ancient coral reefs correspond to three interglacial highstands of sea level 120,000 years ago, 80,000 years ago, and today (Broecker 1965). Similarly, three coral-reef terraces on Barbados match interglacial episodes that occurred 125,000, 105,000, and 82,000 years ago (Broecker *et al.* 1968). Submerged coastal features record lower sea levels during the Quaternary. Examples are the drowned mouths of rivers (rias), submerged coastal dunes, notches and benches cut into submarine slopes, and the remains of forests or peat layers lying below modern sea level. The lowering of the sea was sub-stantial. During the Riss glaciation, the sea fell by some 137–159 m, while during the last glaciation (the Würm) a figure of 105–123 m is likely. A fall of 100 m or thereabouts during the last glaciation was enough to link several islands with nearby mainland: Britain to main-land Europe, Ireland to Britain, New Guinea to Australia, and Japan to China. It would have also led to the floors of the Red Sea and the Persian Gulf becoming dry land.

Sea level rose following the melting of the ice, which started around 12,000 years ago. This speedy rise is the Holocene or Flandrian transgression. It was very speedy at first, up to about 7,000 years ago, and then slowed (Figure 34). Steps on coastal shelves suggest that the rapid transgression involved stillstands, or even small regressions, superimposed on an overall rise. The spread of sea over land during this transgression would have been swift. In the Persian Gulf regions, an advance rate of 100–120 m a year is likely and even in Devon and Cornwall, England, the coastline would have retreated at about 8 m a year.

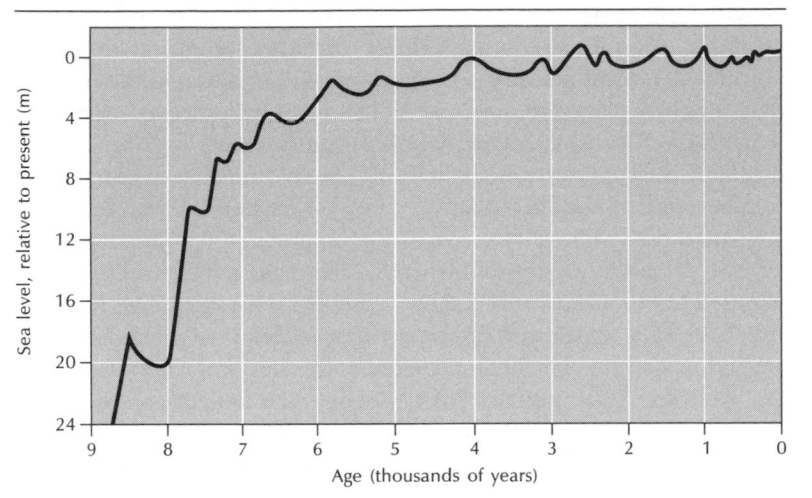

Figure 34 Flandrian transgression in north-western Europe. Notice the rapid rise between about 9,000 and 7,000 years ago; this amounted to about 20 m, an average increase of 1 m per century.

Source: Adapted from Mörner (1980)

Current sea-level rise in response to **global warming** forms the focus of much research, not only on the degree of rise in different parts of the oceans, but also on the impact on coastal populations, **ecosystems**, and landforms. The global average rate of rise is at present unusually fast at about 3 mm per year, which will continue (perhaps at a slower 1.8 mm per year or so) as the oceans warm up and expand and as glaciers and ice caps melt. The rise shows regional variations and will last for several hundred years, even in the best-case scenario.

Further reading: Douglas *et al.* 2001

SOIL

The idea of soil is complicated and soil, like love and home, is difficult to define (Retallack 2003). Geologists and engineers see soil as soft, unconsolidated rock. The entire profile of weathered rock and unconsolidated rock material, of whatever origin, lying above unaltered bedrock is then soil material. By this definition, soil is the same as regolith, that is, 'all surficial material above fresh bedrock' (Ollier and Pain 1996, 2). It includes *in situ* weathered rock (saprolite), disturbed weathered rock (residuum), transported surficial sediments, chemical products, topsoil, and a miscellany of other products, including volcanic ash. Most pedologists regard soil as the portion of the regolith that supports plant life and where soil-forming processes dominate (e.g. Buol *et al.* 2003). This definition poses problems. Some saline soils and laterite (an iron-rich subsoil chiefly found in tropical areas) surfaces cannot support plants – are they true soils? Is a lichen-encrusted bare rock surface a soil? Pedologists cannot agree on these vexatious matters. A possible way of skirting the problem is to define exposed hard rocks as soils (Jenny 1980, 47). This suggestion is not as half-baked as it appears on first acquaintance. Exposed rocks, like soils, are influenced by climate. Like some soils, they will support little or no plant life. Pursuing this idea, soil may be defined as 'rock that has encountered the ecosphere' (Huggett 1995, 12). This definition eschews the somewhat arbitrary distinctions between soil and regolith, and between soil processes and geomorphic processes. It means that the pedosphere is the part of the lithosphere living things affect, and that 'the soil' includes sedimentary material affected by physical and chemical processes, and to far lesser degree, by biological processes. Palaeopedologists (scientists interested in past soils) favour a definition along these lines (e.g. Nikiforoff 1959; Retallack 1990, 9).

If pedologists feel unhappy with a geological definition of soil, then they can use a homegrown pedological term – solum. The solum is the genetic soil developed by soil-building forces (Soil Survey Staff 1999), and normally comprises the A and B horizons of a soil profile, that is, the topsoil and the subsoil. It is the 'soil proper' or edapho-sphere (Huggett 1995, 13). The other portion of the pedosphere, the bit lying beneath the edaphosphere but above the limit of the eco-sphere's influence, is the debrisphere (Huggett 1995, 13). It includes all weathered materials at the Earth's surface and at the bottom of rivers, lakes, and oceans that are virtually unaffected by animals, plants, and microorganisms. It is approximately equivalent to the decomposi-tion sphere (Büdel 1982), but includes detritus created by mechanical disintegration, as well as the productions of chemical weathering. The debrisphere is characterized by epimorphic processes (weathering, leaching, new mineral formation, and inheritance), while the edapho-sphere is characterized by the 'traditional' soil-forming processes (cf. Paton *et al.* 1995, 110). Interestingly, some pedologists are broadening the scope of their purview to include the entire regolith (e.g. Creemans *et al.* 1994).

A soil profile is pedogenetically altered material plus the deep layers (the substrata) that influence **pedogenesis**. The 'deep layers' are called parent material or parent rock, and also lithospheric material (Paton *et al.* 1995, 108). Parent material is material from the lithosphere (igneous, metamorphic, and sedimentary rocks) or biosphere (peat and other organic debris) lying within the influence of, and subject to alteration by, the ecosphere. It is the material in the debrisphere. Parent materials derived from the lithosphere exist in a virtually unaltered state only in those parts of the lithosphere that the biosphere cannot reach. Such little-altered material is grandparent material or bedrock. It is not part of the debrisphere. However, the ecosphere's influence runs deep indeed. Bacterial **populations** exist more than 500 m below the floor of the Pacific Ocean, and microbial life is widespread down to 4,200 m in the continental lithosphere where it may be involved in subterranean geochemical processes – that's a colossal soil profile!

Soil profiles usually show a series of roughly parallel, horizontal, and more-or-less distinct layers known as soil horizons. These horizons are labelled by a system of capital letters, which signify main divisions, and lower case letters, which signify subdivisions. Unfortunately, no lettering system is internationally agreed. As a rule, A and E horizons are depleted in solutes, colloids, and fine particles by the processes of eluviation and pervection (mechanical eluviation of silt and clay), whereas B horizons are enriched in the solutes, colloids, and fine particles lost from the overlying A horizons. C horizons are weathered parent material. O horizons are organic material that accumulates on the soil surface. The A, E, B, and C horizon labels were devised to describe mid-latitude soils. Their applicability to tropical and subtropical soils is debatable. Tropical and subtropical soil horizons are sometimes denoted by the letters M, S, and W (e.g. Williams 1968), which describe the tripartite profile of many tropical and subtropical soils – a mineral layer (M), overlies a stony layer (S), that sits upon weathered rock (W). It now seems likely that mid-latitude A, E, and B–C horizons are equivalent to tropical M, S, and W horizons (Johnson 1994).

Soil profiles differ from one another in varying degrees and may be classified accordingly. Soil classification is essentially a matter of comparing horizon sequences, as well as chemical and physical horizon properties. Schemes of soil classification are multifarious, nationalistic, and use confusingly different nomenclature. Geography and genesis formed the basis older systems, which designated soil orders as zonal, intrazonal, and azonal; divided these into suborders; and then subdivided the suborders in Great Soil Groups such as tundra soils, desert soils, and prairie soils. Newer systems give more emphasis to measurable soil properties that either reflect the genesis of the soil or else affect its **evolution**. The Soil Conservation Service of the US

Department of Agriculture prepared and published the most detailed and comprehensive new classification in 1975 (Soil Survey Staff 1975, 1999). To ease communication between soil surveyors, the nomenclature eschews the early genetic terms and, for units above the series level, uses names derived mainly from Greek and Latin. The **taxonomy** is based on class distinction according to precisely defined diagnostic horizons, soil moisture regimes, and soil temperature regimes. Eleven orders are distinguished – Entisols, Vertisols, Inceptisols, Aridisols, Mollisols, Spodosols, Alfisols, Ultisols, Oxisols, Histosols, and Andisols. The orders successively subdivide into suborders, great groups, subgroups, families, and series.

Further reading: Ashman and Puri 2002; Brady and Weil 2007; Buol *et al.* 2003; White 2005

SOIL–LANDSCAPES

A close association between **soils**, sediments, water, and **topography** is seen in landscapes. Researchers have proposed several frameworks for linking pedological, hydrological, and geomorphic processes within landscapes, most them concerned with two-dimensional **catenas**. The idea of soil–landscape **systems** was an early attempt at an integrated, three-dimensional model (Huggett 1975). The argument was that dispersion of all the debris of weathering – solids, colloids, and solutes – is, in a general and fundamental way, influenced by land-surface form, and organized in three dimensions within a framework dictated by the drainage network. In moving down slopes, weathering products tend to move at right angles to land-surface contours. Flowlines of material converge and diverge according to contour curvature. The pattern of vergency influences the amounts of water, solutes, colloids, and clastic sediments held in store at different landscape positions. Naturally, the movement of weathering products alters the topography, which in turn influences the movement of the weathering products – there is **feedback** between the two systems.

If soil **evolution** involves the change of a three-dimensional mantle of material, it is reasonable to propose that the spatial pattern of many soil properties will reflect the three-dimensional topography of the land surface. This hypothesis can be examined empirically, by observation and statistical analysis, and theoretically, using mathematical models. Early work looked at three-dimensional topographic influences on soil properties in small **drainage basins** (Huggett 1973, 1975; Vreeken 1973), and at soil–slope relationships within a drainage basin (Roy *et al.* 1980). Investigating the effect of landscape setting on **pedogenesis** requires a characterization of topography in three dimensions. The first attempts to describe the three-dimensional character of topography (Aandahl 1948; Troeh 1964) have developed into sophisticated descriptors derived from digital terrain models (e.g. Moore *et al.* 1991). Topographic attributes that appear to be important are those that apply to a two-dimensional catena (elevation, slope, gradient, slope curvature, and slope length) and those pertaining to three-dimensional land form (slope aspect, contour curvature, and specific catchment area, and so on). Research confirms that a three-dimensional topographic influence does exist, and that some soil properties are very sensitive to minor variations in topographic factors. An investigation into natural nitrogen-15 abundance in plants and soils within the Birsay study area, southern Saskatchewan, Canada, showed that, in a small (110 × 110 m) section of an irrigated field of durum wheat, minor topographical variations had a significant influence on landscape patterns of nitrogen-15 in soil and plants (Sutherland *et al.* 1993).

Another model of the soil–landscape continuum built upon technical advances made through linking hydrology and land-surface form for terrain-based modelling of hydrological processes (McSweeney *et al.* 1994). In this new model of soil–landscape systems, geographic information system (GIS) technology integrates various sources of spatial data (e.g. vegetation and geological substrate) and attribute data (e.g. soil organic-matter content and particle-size distribution). The model involves characterizing four things. First is physiographic domain characterization, which brings together available data on geology, climate, vegetation, as well as remotely sensed data, to define and to characterize the physical geography of an area. Second is geomorphometric characterization, which uses digital terrain models to establish landscape topographical attributes. Third is soil horizon characterization, which employs field sampling. Fourth is soil property characterization, which involves the laboratory and statistical analyses of soil horizon attributes collected during the third stage. An underlying assumption in this model is that the same suite of processes link soil and geomorphic patterns. If the assumption should be correct – and a deal of evidence suggests that it is – then in many parts of landscapes, land form should correlate strongly with the underlying nature and arrangement of soil horizons. The challenge is to establish where landform–soil horizon correlations are strong; to determine the feasibility of using such relationships to extrapolate soil patterns across the landscape; and to interpret the relationships through processes and events that shape soil–landscape evolution (McSweeney *et al.* 1994). However, some landscapes contain very complex landform–soil horizon patterns at small scales that require special treatment if they are to be portrayed in a three-dimensional model. Landscapes with cradle–knoll or gilgai microtopography are cases in question. Additionally, deciphering landform–soil horizon relationships in landscapes where subsurface features, such as varied bedrock, or processes, such as saline groundwater, exert a powerful influence on soil patterns may prove especially difficult.

Investigation into soil–landscapes has become highly sophisticated. Researchers study various aspects of soils, hydrology, vegetation, and biogeochemical flows. For instance, a research team is investigating complex spatial patterns of biogeochemical in soils and ecosystems in the Florida Everglades, USA (e.g. Corstanje *et al.* 2006; Grunwald and Reddy 2008).

SOLAR FORCING

The activity of the Sun, and so its emission of radiant **energy** (solar irradiance), varies over days, years, and longer. Before the advent of satellites, proxy variables of solar activity (such as sunspot numbers) had to suffice. Satellites enable researchers to measure total solar irradiance (insolation), which varies about 1.3 W/m^2 (or 0.1 per cent) over an 11-year sunspot cycle. The instrumental and palaeoenvironmental records register longer cycles in solar activity (Table 8). The crucial question concerns the effects that such variations in solar output have on climate through solar forcing.

A connection between the sunspot cycle and Earth surface processes seems undeniable, though the causal links remain elusive. Several hypothesis exist linking **climate change** to changes in the brightness of the Sun (total irradiance), changes in ultraviolet irradiance, changes in the solar wind and the Sun's magnetic flux (which alter the cosmic ray flux), and the effects of cosmic rays on cloud cover.

Periods of prolonged solar minima have occurred during the last several centuries (Figure 35). Some climatologists argue that they tend to be associated with periods of low temperatures (Eddy 1977a, 1977b, 1977c). The Maunder Minimum, for example, coincides with the Little Ice Age, though the correlation between these two events is questionable (e.g. Eddy 1983; Legrand *et al.* 1992). The matter is unresolved. However, recent observational and theoretical studies indicate a definite connection between periods of low solar activity and atmospheric properties. Observations of radiocarbon in tree rings

Table 8 Solar cycles and their possible effects on the environment

Solar cycle	Approximate period (years)	Examples in climatic data
Sunspot or Schwabe	~11	Air temperature records, USA
		Permafrost temperatures and snowfall, northern Alaska
		Annul minimum temperatures, US Gulf Coast region
		Tree-ring widths in many places
Hale	~22	Varve thickness in Elk Lake, Minnesota, and Soppensee, Switzerland
Gleissberg	~87 (70–100)	Northern Hemisphere land and air temperatures over the last 130 years
De Vries– Suess or solar orbital	~210	Air temperatures since Little Ice Age
		Radiocarbon in tree rings over last 9,000 years
Hallstatt or bimillennial	~2,300	Radiocarbon in tree rings over last 9,000 years

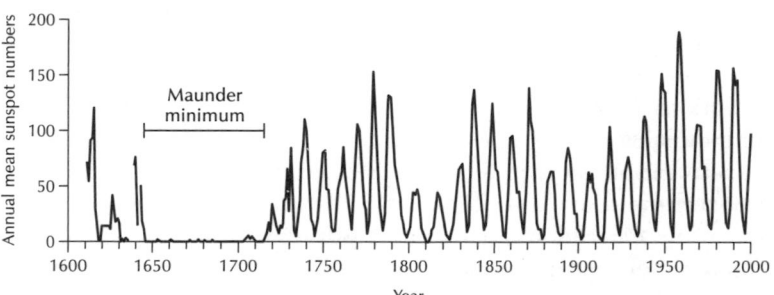

Figure 35 Annual mean sunspot numbers, AD 1610–2000. The period between about AD 1645 and 1715, when the Sun seems to have been quiet and sunspots to have been scarce, is the Maunder Minimum. However, this dearth of sunspots might reflect poor historical records. The 11-year sunspot cycle and the 80-year Gleissberg cycle may be picked out in the post-1800 data.

Source: After Huggett (2007c)

suggest reduced solar activity throughout the Maunder Minimum, and on the theoretical front, a study made with a simple energy-balance climate model demonstrated that cool periods like the Maunder Minimum could have been produced by a 0.22–0.55 per cent reduction in solar irradiance (Wigley and Kelly 1990).

An interesting question is the extent to which current **global warming** relates to solar forcing rather than human activities. A study and review of existing literature by Peter Foukal and his colleagues (2006) concluded that solar luminosity has shown no net increase since the mid-1970s, and that changes in solar output since the seventeenth century are unlikely to have played a major part in global warming. However, they do add the caveat that more subtle influences on climate from cosmic rays or the Sun's ultraviolet radiation may play a role.

Further reading: Soon and Yaskell 2003

SPECIATION

New species arise through the process of speciation. Biologists fiercely debate the nature of speciation and its causes. Key questions centre on mechanisms for bringing new species into existence and mechanisms for maintaining them and building them into cohesive units of inter-breeding individuals that maintain some degree of isolation and indi-viduality. There seems to be a **threshold** at which microevolution (**evolution** through **adaptation** within species) becomes macroevo-lution (evolution of species and higher taxa). Once over this threshold, evolutionary processes act to uphold the species integrity and fine-tune the new species to its niche: gene flow may smother variation; unusual genotypes may be less fertile, or may be eliminated by the **environment**, or may be looked over by would-be mates.

Various mechanisms may thrust a **population** through the specia-tion threshold, each being associated with a different model of specia-tion: allopatric, peripatric, stasipatric, parapatric, and sympatric (Figure 36). Evolutionary biologists argue over the effectiveness of each type of speciation (e.g. Losos and Glor 2003).

Allopatric speciation occurs when geographical isolation reduces or stops gene flow, severing genetic connections between once inter-breeding members of a continuous population. If isolated for long enough, the two daughter populations will probably evolve into differ-ent species. This mechanism is the basis of the classic model of allopatric ('other place' or geographically separate) speciation, as propounded by Ernst Mayr (1942) who called it geographical speciation and saw geographical subdivision as its driving force. Mayr recognized three kinds of allopatric speciation: strict allopatry without a population bottleneck, where a population extends its range over a barrier and the two subpopulations then evolve independently; strict allopatry with a

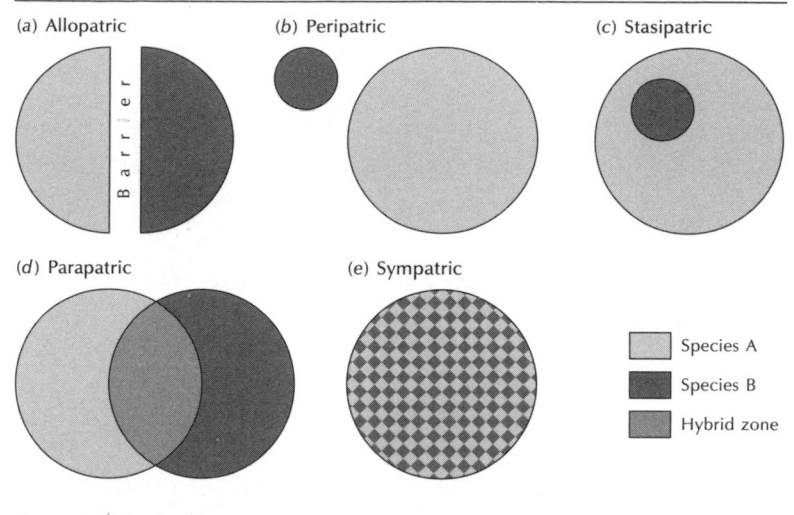

Figure 36 Types of speciation.

Source: After Huggett (2004)

population bottleneck, where a few founder individuals from a population colonize a new area; and **extinction** of intermediate populations in a chain of races. The last case applies to subspecies of *Ensatina*
salamanders in California, which display subtle morphological and
genetic differences all along their large ring-shaped range around
Central Valley. They all interbreed with their immediate neighbours,
save where the extreme ends of the range overlap in southern California – the large-blotched ensatina (*E. eschscholtzii klauberi*) and the
Monterey ensatina (*E. e. eschscholtzii*) do not interbreed. **Vicariance**
events and **dispersal**-cum-founder events may drive allopatric speciation (Figure 37). Two species of North American pines illustrate
vicariance speciation. Western North American lodgepole pine (*Pinus
contorta*) and eastern North American jack pine (*Pinus banksiana*)
evolved from a common ancestral population that the advancing
Laurentide Ice Sheet split asunder some 500,000 years ago. The colonization of the Galápagos archipelago from South America by an
ancestor of the present giant tortoises (*Geochelone* spp.), probably
something like its nearest living relative the Chaco tortoise (*Geochelone
chilensis*), is an example of a dispersal and founder event.

Peripatric speciation is a subset of allopatric speciation. It occurs in
populations on the edge (perimeter) of a species range that become
isolated and evolve divergently to create new species. A small founding
population is often involved. An excellent example of this is the paradise kingfishers (*Tanysiptera*) of New Guinea (Mayr 1942). The main

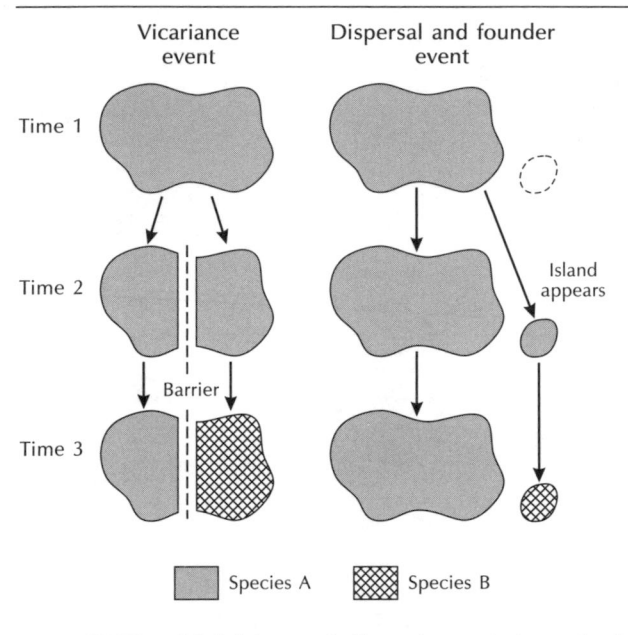

Figure 37 The chief drivers of allopatric speciation: vicariance events and dispersal–cum–founder events.

Source: Adapted from Brown and Lomolino (1998)

species, the common paradise kingfisher (*Tanysiptera galatea galatea*), lives on the main island. The surrounding coastal areas and islands house a legion of morphologically distinct races of the paradise king-fishers.

Parapatric (abutting) speciation is the outcome of divergent evolution in two populations living geographically next to each other. The divergence occurs because local **adaptations** create genetic gradients or clines. Once established, a cline may reduce gene flow, especially if the species is a poor disperser, and selection tends to weed out hybrids and increasingly pure types that wander and find themselves at the wrong end of the cline. A true hybrid zone may develop, which in some cases, once reproductive isolation is effective, will disappear to leave two adjacent species. An example is the main species of the house mouse in Europe, in which a zone of hybridization separates the light–bellied eastern house mouse (*Mus musculus*) of eastern Europe and the dark–bellied western house mouse (*Mus domesticus*) of western Europe (Hunt and Selander 1973).

Sympatric speciation involves genetic divergence without geo-graphical isolation. It occurs within a single geographical area and the

new species overlap – there is no spatial separation of the parent population. Separate genotypes evolve and persist while in contact with
each other. Once deemed rather uncommon, new studies suggest that
sympatric speciation may be a potent process of evolution (e.g. Via
2001) that happens in the presence of gene flow (Nosil 2008;
Niemiller *et al.* 2008). Several processes appear to contribute to
sympatric speciation. Disruptive selection favours extreme phenotypes
and eliminates intermediate ones. Once established, **natural selection** encourages reproductive isolation through **habitat** selection or
positive assertive mating (different phenotypes choose to mate with
their own kind). Habitat selection in insects may have favoured
sympatric speciation and account for much of the large diversity of
that group. Competitive selection, a variant of disruptive selection,
favours phenotypes within a species that avoid intense competition
and clears out intermediate types. Two species of palms (*Howea*)
growing on remote Lord Howe Island seem to have evolved sympat-
rically (Savolainen *et al.* 2006).

Stasipatric speciation is speciation within a species range owing to
chromosomal changes. Chromosomal changes occur through: (1) a
change in chromosome numbers, or (2) a rearrangement of genetic
material on a chromosome (an inversion) or a transferral of some
genetic material to another chromosome (a translocation). Polyploidy
increases twofold or more the normal chromosome component, and
polyploids are often larger and more productive than their progenitors. Polyploidy is rare in animals but appears to be a major source of
sympatric speciation in plants: 43 per cent of dicotyledon species and
58 per cent of monocotyledon species are polyploids. Stasipatric
speciation seems to have occurred in some western house mouse (*Mus
domesticus*) populations. In Europe, the normal karyotype for the
species contains 20 sets of chromosomes. Specimens with 13 sets of
chromosomes were first discovered in south-east Switzerland in the
Valle di Poschiavo. At first, these were classed as a new species and
designated *Mus poschiavinus*, the tobacco mouse. Later, specimens from
other alpine areas of Switzerland and Italy (as well as from North
Africa and South America) also had non-standard karyotypes.
Surprisingly, all the populations showed no morphological or genetic
differences other than differences in their karyotypes and all belonged
to *Mus domesticus*.

Further reading: Coyne and Orr 2004; Dieckmann *et al.* 2004

SUCCESSION

The idea of vegetation (or ecological) succession has proved hugely important in ecology. Henry Chandler Cowles (1899) proposed a formal concept of succession based on the vegetation of sand dunes (the Indiana Dunes) on the shores of Lake Michigan, USA. Frederic Edward Clements (1916) developed the idea of succession, which he saw as a predetermined sequence of developmental stages, or sere, that ultimately leads to a self-perpetuating, stable community called climatic climax vegetation that then endures. He recognized six stages in any successional sequence: (1) nudation – an area is left bare after a major **disturbance**; (2) migration – species arrive as seeds, spores, and so on; (3) ecesis – the plant seeds establish themselves; (4) competition – the established plants complete with one another for resources; (5) reaction – the established plants alter their **environment** and so enable other new species to arrive and establish themselves; (6) stabilization – after several waves of colonization, an enduring **equilibrium** is achieved. Clements distinguished between primary succession and secondary succession. Primary succession occurs on newly uncovered bare ground that has not supported vegetation before. New oceanic islands, ablation zones in front of glaciers, developing sand dunes, fresh river alluvium, newly exposed rock produced by faulting or volcanic activity, and such human-made features as spoil heaps are all open to first-time colonization. The full sequence of successional communities forms a primary sere (prisere), with different priseres occurring on different substrates: hydrosere in open water (where succession could lead to woodland); haloseres in salt marshes; psammosere on sand dunes; and lithoseres on bare rock. Secondary succession occurs on severely disturbed ground that previously supported vegetation. Fire, flood, forest clearance, the removal of grazing animals, hurricanes, and many other factors may inaugurate secondary succession.

Facilitation, tolerance, and inhibition are the three mechanisms thought to drive succession (Horn 1981). Researchers disagree about the relative importance of these mechanisms, so there are three main models of succession. First is the facilitation model. The argument here is that pioneer species make the **habitat** less suitable for themselves and more suitable for a new round of colonists. The process of reaction continues, each group of species facilitating the colonization of the next group. This is the classical Clementsian model of succession. Reaction helps to drive successional changes. On a sandy British beach, the first plant to colonize is marram-grass (*Ammophila arenaria*). After a rhizome fragment takes hold, it produces aerial shoots. The shoots impede wind flow and sand tends to accumulate around them. The sand gradually buries the plant, which, to avoid 'suffocation', grows longer shoots. The shoots keep growing and the mound of sand keeps growing. Eventually, a sand dune forms. This is then colonized by other species, including sand fescue (*Festuca rubra* var. *arenaria*), sand sedge (*Carex arenaria*), sea convolvulus (*Calystegia soldanella*), and two sea spurges (*Euphorbia paralias* and *Euphorbia portlandica*), which help to stabilize the sand surface. Second, the tolerance model argues that late successional plants, as well as early successional plants, may invade in the initial stages of colonization. In northern temperate forests, for example, late successional species appear almost as soon as early successional species in vacant fields. The early successional plants grow faster and soon become dominant. However, late successional plants maintain a foothold and come to dominate later, crowding out the early successional species. In this model, succession is a thinning out of species originally present, rather than an invasion by later species on ground prepared by specific pioneers. Any species may colonize at the outset, but some species are able to outcompete others and come to dominate the mature community. Third, the inhibition model takes account of chronic and patchy disturbance, a process that occurs when, for example, strong winds topple trees and create forest gaps. Any species may invade the gap opened up by the toppling of any other species. Succession in this case is a race for uncontested dominance in recent gaps, rather than direct competitive interference. No species is competitively superior to any other. Succession works on a 'first come, first served' basis – the species that happen to arrive first become established. It is a disorderly process, in which any directional changes are due to short-lived species replacing long-lived species. This model has much in common with the notion of multidirectional or multiple pathway succession (see **community change**).

Causes within a community and causes outside a community may drive succession. The members of a community propel autogenic (self-generated) succession. In the facilitation model, autogenic succession is a unidirectional sequence of community, and related **ecosystem**, changes that follow the opening up of a new habitat. The sequence of events takes place even where the physical environment is unchanging. An example is the heath cycle in Scotland (Watt 1947), where heather (*Calluna vulgaris*) is the dominant heathland plant. As a heather plant ages, it loses its vigour and is invaded by lichens (*Cladonia* spp.). In time, the lichen mat dies, leaving bare ground, which is in turn invaded by bearberry (*Arctostaphylos uva-ursi*), which eventually succumbs to invasion by heather. The cycle takes about 20–30 years.

Fluctuations and directional changes in the physical environment steer allogenic (externally generated) succession. A host of environmental factors may disturb communities and ecosystems by disrupting the interactions between individuals and species. For example, deposition occurs when a stream carries silt into lake. Slowly, the lake may change into a marsh or bog, and eventually the marsh may become dry land.

An understanding of succession is immensely important in guiding the restoration of ecosystems. Restoration ecology is a growing discipline that draws upon studies of natural succession and that, in its turn, provides insights into successional mechanisms (e.g. Walker *et al.* 2007).

Further reading: Dale *et al.* 2005; Walker and Moral 2003; Walker *et al.* 2007

SUSTAINABILITY

The Siamese twin ideas of 'sustainability' and 'sustainable development' emerged in the 1990s as a focus for the shared needs of society, the economy, and the **environment**. They gained rapid momentum and acceptance and remain a dominant development philosophy in the opening decade of the twenty-first century. Their conceptual role in physical geography is debatable, but they at least form the focus for some physical geographical research.

As with many buzzphrases, it is tricky to find a wholly satisfactory definition of sustainable development. In 1987, The Brundtland Report *Our Common Future* offered the most familiar definition of sustainable development: 'meeting the needs of the present without compromising the ability of future generations to meet their own needs' (WCED 1987, 8). Sustainable development breaks down into three parts – environmental sustainability, economic sustainability, and socio-political sustainability. During the development and adoption of the idea of 'sustainable development', commentators pointed to its general ambiguity and vagueness as a strength and as a weakness. However, despite the rhetoric surrounding the term sustainable development, its success in focusing environmental philosophy and practice is indisputable, especially since the UN General Assembly's follow up of a recommendation in the Bruntland Report that culminated in the 1992 Earth Summit and Global Forum. From this conference came two important things: conventions on climate and **biodiversity**; and Agenda 21 (the Rio declaration in Environment and Development), which was a detailed plan to translate the guiding principle of sustainability into a reality.

A rich focus of research on the biophysical environment is resource consumption and flows of waste products and pollutants that affect the natural environment and human society. Such research is in part the subject matter of industrial ecology, which has its own journal. Processes associated with resource consumption and resource flows operate on different spatial scales, from local to global, and exhibit high rates of change that make it difficult for living things to adapt. This applies, for instance, to animals (including humans) and plants adapting to rapid **climate change** and other human-induced changes. Several key ideas, which are highly pertinent to physical geography, have emerged from such research. These ideas include ecological footprint, geomorphic footprint, and **carrying capacity**. The ecological footprint is the land and water area a human population requires to produce the resources it consumes and to absorb the waste products of consumption, under prevailing technology. For instance, in 2003, Malawi's ecological footprint was 0.6 global hectares per person, the United Kingdom's was 5.6, and the world's was 2.2, which is about 23 per cent more than the world can regenerate (Global Footprint Network). The 'geomorphic footprint' is a measure of the rate at which humans create new landforms and mobilize sediment (Rivas *et al.* 2006). 'Footprints' are proliferating and include the much-vaunted carbon footprint.

Further reading: Rees 1995; WCED 1987; www.footprintnetwork.org

SYSTEMS

Many physical geographers adopt a systems approach to their subject. A perusal of the literature reveals **ecosystems, drainage basin** systems, hydrological systems, fluvial systems, meteorological systems, **soil** systems, and many more. It is perhaps easiest to comprehend what a system is by taking an example – a hillslope system will suffice. A hillslope extends from an interfluve crest, along a valley side, to a sloping valley floor. It is a system insofar as it consists of things (rock waste, organic matter, and so forth) arranged in a particular way. The arrangement is seemingly meaningful, rather than haphazard, because it is explicable in terms of physical processes (Figure 38). The 'things' of which a hillslope is composed may be described by such variables as particle size, soil moisture content, vegetation cover, and slope angle. These variables, and many others, interact to form a regular and connected whole: a hillslope, and the mantle of debris on it, records a propensity towards reciprocal adjustment among a complex set of

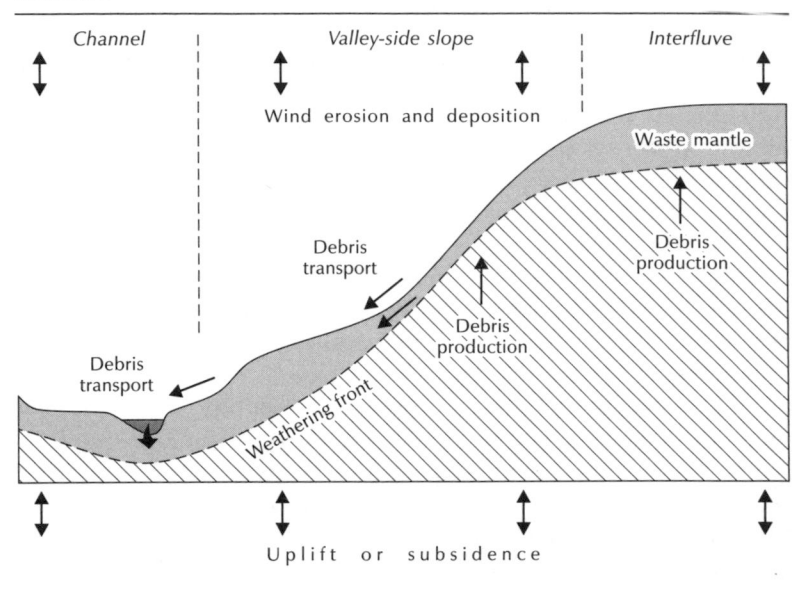

Figure 38 A hillslope as a system, showing storages (waste mantle), inputs (e.g. wind deposition and debris production), outputs (e.g. wind erosion), throughputs (debris transport), and units (channel, valley-side slope, interfluve).

Source: After Huggett (2007a)

variables. The complex set of variables includes rock type, climate, tectonic activity, and hillslope form. Rock type influences weathering rates, the geotechnical properties of the soil, and rates of infiltration. Climate influences slope hydrology and so the routing of water over and through the hillslope mantle. Tectonic activity may alter base level. Hillslope form, acting mainly through slope angle and distance from the divide, influences the rates of processes such as landsliding, creep, solifluction, and wash. Change in any of the variables will tend to cause a readjustment of hillslope form and process.

Systems of all kinds are open, closed, or isolated according to how they interact, or do not interact, with their surroundings (Huggett 1985, 5–7). Traditionally, an isolated system is a system that is completely cut off from its surroundings and that cannot therefore import or export matter or **energy**. A closed system has boundaries open to the passage of energy but not of matter. An open system has boundaries across which energy and materials may move. All systems in physical geography, including hillslopes, are open systems as they exchange energy and matter with their surroundings. In addition, they all have internal and external variables. Take a drainage basin. Soil wetness, streamflow, and other variables lying inside the system are endogenous or internal variables. Precipitation, solar radiation, tectonic uplift, and other such variables originating outside the system and affecting drainage basin dynamics are exogenous or external variables.

It is vital to stress that systems are mental constructs and scientists define them in various ways. Two conceptions of systems are important in physical geography: systems as process and form structures and systems as simple and complex structures (Huggett 1985, 4–5, 17–44). Physical geographers recognize three types of process and form systems: form systems, process systems, and form and process systems. Form or morphological systems are sets of form variables believed to interrelate in a meaningful way in terms of system origin or system function. The form of a hillslope system is describable using several measures. Form elements would include measurements of anything on a hillslope that has size, shape, or physical properties. A simple characterization of hillslope form is shown in Figure 39a, which depicts a cliff with a talus slope at its base. This 'form system' simply shows that the talus lies below the cliff; no causal connections between the processes linking the cliff and talus slope are inferred. Digital terrain models enable the making of sophisticated characterizations of hillslope and land-surface forms. Process systems (cascading or flow systems) are 'interconnected pathways of transport of energy or matter or both, together with such storages of energy and matter as may be required' (Strahler 1980, 10). An example is a hillslope represented as a store of materials: weathering of bedrock and wind deposition adds materials to the store, and erosion by wind and fluvial erosion at the slope base removes materials from the store. The materials pass through the system and in doing so link the morphological components. In the case of the cliff and talus slope, it could be assumed that rocks and debris fall from the cliff and deliver energy and rock debris to the talus below (Figure 39b). Process–form systems (process–response systems) comprise an energy-flow system linked to a form system in such a way that system processes may alter the system form and, in turn, the changed system form alters the system processes. A hillslope viewed in this way has slope form variables interacting with slope process variables. In the cliff-and-talus example, rock falling off the cliff builds up the talus store (Figure 39c). However, as the talus store increases in size, so it begins to bury the cliff face, reducing the area that supplies debris. In consequence, the rate of talus growth diminishes and the system changes at an ever-decreasing rate. The

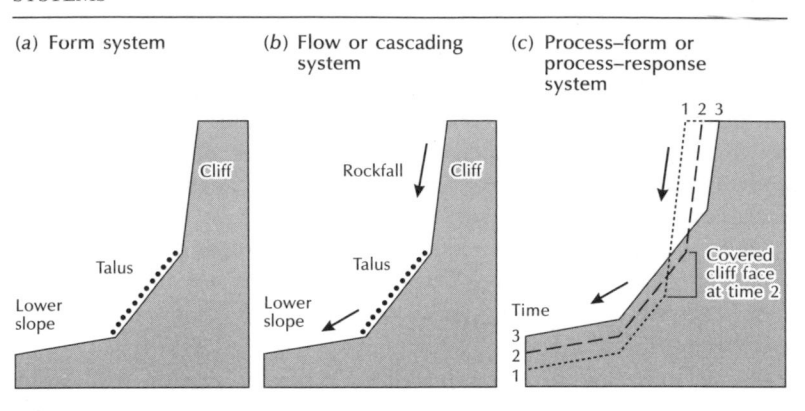

(a) Form system (b) Flow or cascading system (c) Process–form or process–response system

Figure 39 A cliff and talus slope viewed as (a) a form system, (b) a flow or cascading system, and (c) a process–form or process–response system.

Source: After Huggett (2007a)

process described is an example of negative **feedback**, which is an important facet of many process–form systems.

A systems approach sometimes has grander aspirations than the recognition of individual systems and attempts to apply general systems theory (GST), and its offshoot systems theory, to the objects of interest in various physical geographical disciplines. Systems theory adopts a holistic approach and looks at properties common to all systems. For example, systems ecology takes a holistic approach to studying ecosystems, stressing the notion that an ecosystem is a complex system (a system with many interacting parts) with emergent properties (such as vegetation **succession**) that are unpredictable from knowledge of the component parts of the system. Interestingly, emergent properties arise in dissipative systems, in which symmetry breaks down spontaneously (producing anisotropy) and complex, sometimes chaotic, dissipative structures form by the often nonlinear interaction of systems components (see **complexity**). Such structures exist in any open system characterized by irreversible processes. In geomorphology, microscale events of sediment transport and deterministic channel hydraulics cannot predict the behaviour of braided channel switching, which is an emergent property (Harrison 2001). In ecology, emergent properties appear when the community is the focus of attention – the dynamics of a community is more than the sum of its component species, being rather the sum of its species' interactions. It is important to distinguish emergent properties from collective properties, such as species diversity and community biomass.

Many subfields of physical geography and geosciences in general have started adopting a holistic and ever more interdisciplinary view of their systems of interest. Indeed, the term 'Earth System Science' (ESS), which is 'the study of the Earth as a single, integrated physical and social system' (Pitman 2005, 138), has begun to seep into the physical geographical literature in relation to research approaches adopting such a standpoint (e.g. Tooth 2008).

Further reading: Kump *et al.* 2004

TAXONOMY

Taxonomy is the practice and science of classification. It is as important to the physical geographical sciences as it is to any science. It involves recognizing classes (groups of things or situations with one or more shared characteristics) and assigning the properties of the class to a thing or situation. Classes might be volcanoes, slope elements, or plants. Assigning a thing or object to the class 'volcano' means that it has properties characteristic of volcanoes. Taxonomies, or taxonomic schemes, comprise classes or taxonomic units (taxa; singular taxon) often arranged in a hierarchical structure with tiers typically related by subtype–supertype relationships, in which the subtype of class has by definition the same constraints as the supertype class plus one or more additional characteristics. For example, lemur is a subtype of primate, so any lemur is a primate, but not every primate is a lemur – a thing needs to satisfy more constraints to be a lemur than to be a primate.

Biological classification is a method for grouping and categorizing species of organisms. Its modern roots lie in the work of Carl Linnaeus, who grouped species according to shared physical characteristics. Biologists have since revised these groupings to improve consistency with the Darwinian principle of descent from a common ancestor. Molecular systematics uses DNA sequences as the basis for classification. It has led to many recent revisions of the biological hierarchy (e.g. Hackett *et al.* 2008). In phylogenetic taxonomy (or cladistic taxonomy), organisms are classified by clades, which are based on evolutionary grouping by ancestral traits. By using clades as the criteria for separation, cladistic taxonomy, using cladograms, can categorize taxa into unranked groups that give no indication of change through time or rate of **evolution**. The estimation of evolutionary rates leads to phylograms and chronograms. In phylograms, the branch lengths are measured as the number of character changes (substitutions in

DNA sequences) along branches; in chronograms (ultrametric trees), they are measured in absolute time (see Figure 3 for an example).

Numerical taxonomy uses cluster analysis (the solving or best-fitting of numerical equations that characterize all measurable quantities of a set of objects). It was once relatively popular in some physical geographical sciences, including pedology and ecology, and it still finds currency in palynology and elsewhere.

TECTONICS AND NEOTECTONICS

Tectonics is study of the form and **evolution** of the Earth's major crustal structures – mountain ranges, plateaux, fold belts, islands arcs, and so on. Of prime concern are the forces and movements that build mountains, fold rocks, and crack and deform cratons. Tectonic studies are also important for understanding erosion patterns in geomorphology, their effects on climate, and as guides for the economic geologist searching for petroleum and metallic ores. Neotectonics, a subfield of tectonics, focuses upon crustal structures formed in geologically recent times, especially those produced by earthquakes. Some geologists take 'geologically recent' to mean currently active, while others push the boundary of the neotectonic period back about 10 million years to the middle Miocene epoch. Since the 1960s, **plate tectonics** has become by far the dominant theory to explain the origin and forces responsible for the tectonic features of the continents and ocean basins.

Further reading: Willett et al. (2006)

TELECONNECTIONS

Teleconnections are links between environmental events, and particularly between climatic variations, separated in time and geographically. They result from the rich interconnectedness of various components in the Earth–atmosphere system. An outcome of such interconnectivity is that a change in one component leads to changes in others, the changes usually taking a while to work through the system and affecting areas remote from the source of the original change. Statistical analysis of climatic data reveals teleconnections, but an understanding of their causes demands the application of atmosphere–ocean general circulation models.

The El Niño–Southern Oscillation (ENSO) is the strongest natural fluctuation of climate occurring over a few years (interannual timescales). The Southern Oscillation, discovered by Sir Gilbert Thomas Walker in the 1923, is the seesaw shift from high to low pressure (and low to high rainfall) in the Pacific and Indian Oceans. The sea-level pressure difference between Tahiti and Darwin – the Southern Oscillation Index (SOI) – measures the state of the Southern Oscillation. El Niño is the appearance every few years of exceptionally warm waters off Ecuador and Peru that extend far westwards in equatorial regions, so reducing the difference in sea-surface temperatures across the tropical Pacific. El Niño events are associated with an air-pressure fall over much of the south-eastern Pacific Ocean, and an air-pressure rise over Indonesia and northern Australia. In consequence of these changes in atmospheric pressure, the trade winds weaken, the Southern Oscillation Index becomes highly negative, and sea level falls in the west Pacific and rises in the east Pacific by up to 25 cm as warm equatorial waters spread eastwards. At the same time, the weakened trade winds lessen the upwelling of cold water in the eastern equatorial Pacific, so reinforcing the initial positive temperature anomaly. The weakened trades also lead to changes in subsurface ocean waters that eventually reduce the sea-surface temperatures off the South American coast. In combination, the tropical air–sea instability and the delayed negative **feedback** due to subsurface ocean dynamics tend to excite oscillations. When cold water returns to the western South American seaboard during a La Niña event, the pressure gradient reverses over the Pacific and Indian Oceans. The combined oceanic and atmospheric changes are the El Niño–Southern Oscillation.

El Niño events occur on average every five years, but the interval between them is irregular, ranging from 2 to 8 years. ENSO events originate in the tropics, and have thoroughgoing impacts on climates within ENSO's equatorial central-eastern Pacific 'source region'; by disrupting the tropical atmospheric circulation; they also affect climates in remote tropical and extratropical regions, possibly as remote as Europe, making ENSO a worldwide phenomenon. The mechanisms by which ENSO events produce distant climatic responses are not fully resolved.

The effects of El Niño and La Niña events are far reaching, but depend on the time of year (Figure 40) (see Glantz 2001; Glantz *et al.* 1991). The climatic changes have had material impacts on **ecosystems**. Reduced coastal upwelling off Peru and Ecuador, coupled with increased rainfall, leads to reduced levels of nutrients and increased concentrations of sediments in coastal waters during long-lasting El Niño events. These **environmental changes** can lead to a collapse of

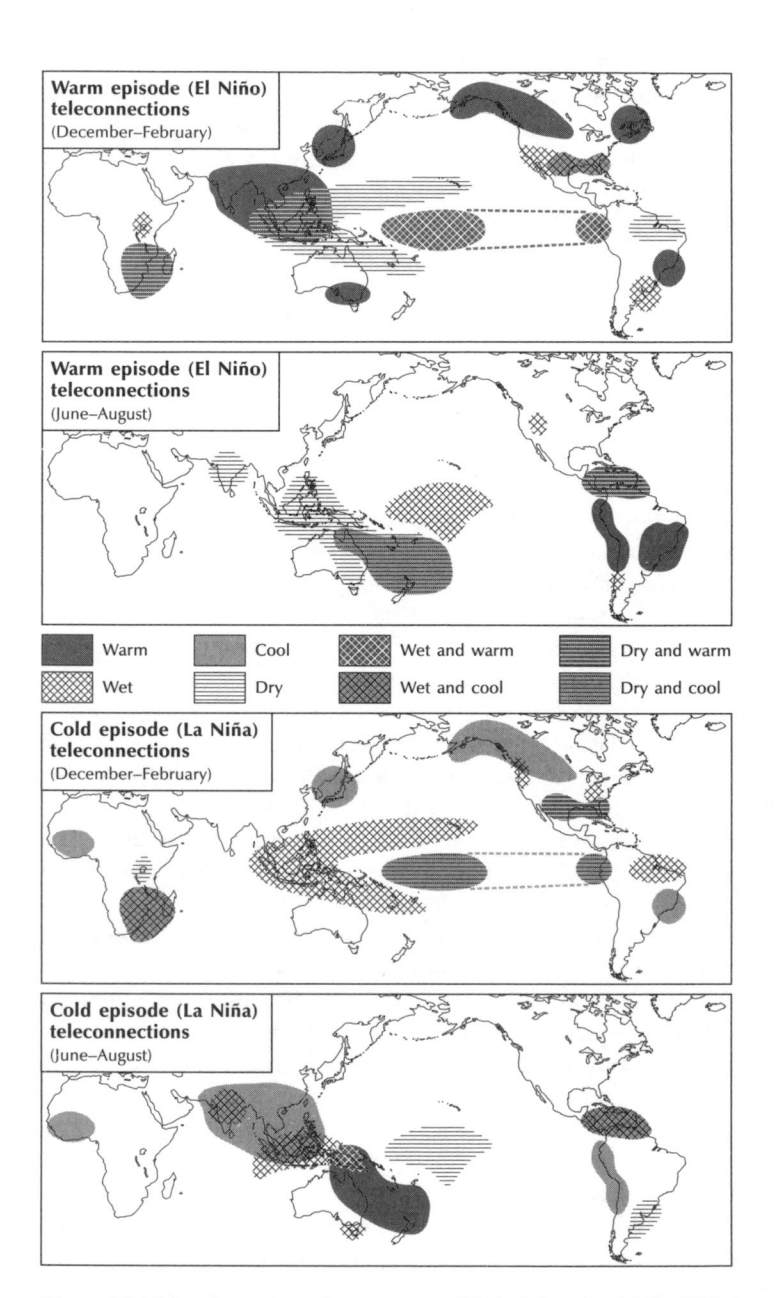

Figure 40 Teleconnections during warm (El Niño) and cold (La Niña) episodes for December to February and June to August.

Source: Adapted from http://en.wikipedia.org/wiki/El_Ni%C3%B1o–Southern_Oscillation

fish **populations**, as happened in 1972 and 1982–83. ENSO droughts may help to explain extensive tropical fires in Brazil and south-east Asia, as occurred in 1997–98. The 1997–98 El Niño event led to the death of around 16 per cent of the world's reef ecosystems, and since that time, mass coral bleaching has become common around the world, with all regions suffering some 'severe bleaching' (Marshall and Schuttenberg 2006). ENSO events may also affect human health, with the incidence of malaria, typhoid, and cholera seemingly increased during ENSO events and famine increased owing to crop failures (WHO 1997). The poor crop yields of 1788–89 in Europe, which resulted from the unusually strong El Niño event of 1789–93, may have assisted the French Revolution (Grove 1998).

Other teleconnections stem from the Pacific–North American Pattern (PNAP), which is partly influenced by changes in ENSO, and the North Atlantic Oscillation (NAO), which may be the regional manifestation of the Arctic Oscillation (AO), a yearly shift of atmospheric mass between the polar cap and the middle latitudes in both the Atlantic and Pacific Oceans. The NAO is the dominant mode of winter climate variability in the North Atlantic region, ranging from central North America to Europe and into northern Asia. It is a large-scale seesaw in atmospheric mass between the Azores subtropical high and the Icelandic polar low. Its index varies from year to year, but tends to stay in one phase for several years. The PNAP and NAO account for a substantial part of the pattern of Northern Hemispheric temperature change, especially in winter. A similar feature – the Southern Annual Mode or Antarctic Oscillation (AAO) is dominant in the Southern Hemisphere. The tropical Indian Ocean, too, has dominant large-scale patterns of ocean–atmosphere interactions. These have characteristics similar to El Niño and relate to large east–west changes in sea-surface temperature and a switch of the major tropical **convection** areas from Africa to Indonesia.

THRESHOLDS

A threshold separates different states of a system. It marks some kind of transition or 'tipping point' in the behaviour, operation, or state of a system. Everyday examples abound. Water in a boiling kettle crosses a temperature threshold in changing from a liquid to a gas. Similarly, ice taken out of a refrigerator and placed upon a table in a room with an air temperature of 10°C will melt because a temperature threshold has been crossed. In both examples, the huge differences in state – solid water to liquid water, and liquid water to water vapour – may result from tiny changes of temperature.

Many geomorphic processes operate only after the crossing of a threshold. Landslides, for instance, require a critical slope angle, all other factors being constant, before they occur. Stanley A. Schumm (1979) made a powerful distinction between external and internal system thresholds. A geomorphic system will not cross an external threshold unless forced to do so by a change in an external variable. A prime example is the response of a geomorphic system to **climate change**. Climate is the external variable. If, say, runoff were to increase beyond a critical level, then the geomorphic system might suddenly respond by reorganizing itself into a new state. No change in an external variable is required for a geomorphic system to cross an internal threshold. Rather, some chance fluctuation in an internal variable within a geomorphic system may take a system across an internal threshold and lead to its reorganization. This appears to happen in some river channels where an initial disturbance by, say, overgrazing in the river catchment triggers a complex response in the river channel: a complicated pattern of erosion and deposition occurs with phases of alluviation and downcutting taking place concurrently in different parts of the channel system.

In **soils**, autogenic (self-generated) changes often occur when internal thresholds are crossed and may be distinguished from allogenic (externally driven) changes. Under Mediterranean climates, a soil threshold is crossed in moving from Alfisols on summits and back-slopes to Vertisols on toeslopes (Muhs 1982). The Vertisols evolve when sufficient smectite (an expanding clay) accumulates, the smectite coming from *in situ* weathering and the downslope movement of suspended clay particles and base cations. There are many other examples of intrinsic soil thresholds (e.g. Muhs 1984). The leaching of carbonates is a prerequisite for clay migration in soils. This is because divalent cations of calcium and magnesium, among others, are strong flocculants that inhibit pervection (mechanical transport) of clay. Conversely, monovalent sodium ions tend to disperse clays. In coastal or playa-margin environments, clay migration occurs rapidly once enough sodium has accumulated to cause clay dispersal. Laminar calcretes only form in arid and semiarid regions when calcium carbonate plugs the pores in growing K horizons. After that time, a laminar zone of calcium carbonate accumulation develops and grows upwards.

Thresholds occur in ecological systems (Groffman *et al.* 2006). They occur where a small change in an internal or external **ecosystem** variable separates abrupt changes in biological, chemical, and physical properties, which may or may not be reversible. A prime example comes from Florida Bay, a 2,200-km^2 shallow estuary at the southern tip of Florida. This ecosystem changed abruptly in the early 1990s from an oligotrophic (low nutrient) clear water system with primary production dominated by sea-grasses (rooted aquatic plants) to a more turbid system with primary production dominated by phytoplankton blooms (Gunderson and Holling 2002). In this case, as in many others, there was a nonlinear response in ecosystem state to **environmental change**, with several variables (water clarity, primary production, nutrient cycling, food webs) changing dramatically once the ecosystem crossed a threshold. The possible causes of the threshold crossing included nutrient input from septic systems, sea-level change, a lack of hurricanes, drought, water diversions, and the removal of grazers. As the Florida Bay case suggests, ecological thresholds have implications for conservation (A. Huggett 2005).

In the Earth climate system, several components are possibly 'tipping elements' that, should their thresholds or tipping points be crossed, lead to abrupt **climate change**. They include boreal forest dieback, dieback of the Amazon forest, change in the frequency or amplitude of ENSO events, a shift in the West African monsoon, change in Atlantic deep-water formation, the loss of ice in the Arctic, and the instability of the West Antarctic ice sheet (Lenton *et al.* 2008).

Further reading: Groffman *et al.* 2006; Phillips 2001

TIME

The role of time in the physical geographical sciences has stirred debate. Geomorphologists and ecologists are perhaps the most involved in the debate, but it has relevance to researchers in all other fields. The crux of the matter concerns the distinction between what happens over relatively short timescales (days to centuries) and what happens over geological timescales (thousands and millions of years).

In ecology, Larry B. Slobodkin (1961) drew a distinction between ecological time and evolutionary time that for over 40 years has structured the mindset of ecologists exploring the processes underlying patterns in nature. He defined ecological time as about 10 generations (a period over which he expected that **populations** could maintain an approximate steady state), and evolutionary time as about half a million years (time enough for evolutionary change to disrupt ecological steady states). In Slobodkin's terms, evolutionary and ecological time differ by orders of magnitude. Although useful in many ways, the distinction Slobodkin drew has been misleading when interpreted too strictly because it perhaps overemphasizes differences between rates of ecological and evolutionary change. Even during the decade of Slobodkin's (1961) publication, several studies reported quite rapid **evolution** in ecologically important phenotypes, and in the 1980s and 1990s, a flood of such examples appeared (see Thompson 1998; Palumbi 2001; Hairston *et al.* 2005). Philip D. Gingerich (2001), drawing on Theodosius Dobzhansky's (1937) distinction between micro- and macroevolution, suggested three evolutionary timescales: generational, microevolutionary, and macroevolutionary. The former two cover the timeframe of ecological change envisioned by Slobodkin (1961) and more recently by Thompson (1998). However, a current view is that ecological time and evolutionary time are insep-arable because evolution takes place much faster than was once thought (Hairston *et al.* 2005).

In geomorphology, arguments surrounding time have tended to focus upon the dichotomy between timebound and timeless aspects of change, that is, in essence, between process studies on the one hand and historical work on the other. These arguments have some simi-larity to the arguments in ecology between ecological and evolu-tionary timescales. The historical school, championed by William Morris Davies among others, chiefly considers changes in landforms over long periods of time – a 'what happened' approach that provides geohistorical knowledge. Arthur N. Strahler (1952) started the process school, which has enjoyed huge popularity over the last 50 years (although its roots pre-date Strahler), evincing an interest in 'what happens' (timeless knowledge) rather than 'what happened' (time-bound knowledge). Put another way, Strahler's approach focuses on 'the dynamic interaction between process and form that leads to a time-independent, self-regulatory steady state in which form remains constant despite continuous fluxes of material and energy through an open geomorphic system', and not on 'the historical development of the form of a particular landscape' (Rhoads 2006, 15). Despite its apparent success, the process approach has problems, including its reliance on reductionism, its inability to deal with complex and large-scale geomorphic phenomena, and its seeming disregard for historical events. Geomorphologists have never adequately bridged the gaping gulf between the timeless and timebound approaches. An early attempt to do so subdivided time according to the independence, semi-dependence, and dependence of variables in landscape evolu-tion: steady time (days), graded time (centuries), and cyclic time (millions of years) (Schumm and Lichty 1965). On short timescales, steady-state ideas may be appropriate; longer timescales may be suitable for dynamic **equilibrium** or cyclic time; the longest geomorphic timescales may encompass **evolutionary geomorphology**. A recent attempt by Bruce L. Rhoads (2006) to meld the process and historical approaches appealed to process philosophy, and especially to the process philosophy developed by Arthur North Whitehead, the British mathematician and philosopher (Rhoads 2006). To understand what process philosophy is, it may help to consider these quotations from Whitehead's canon as picked out by Rhoads: 'Every scheme for the analysis of nature has to face these two facts, *change* and *endurance*. . . . The mountain endures. But when after the ages it has been worn away, it has gone' (Whitehead 1925, 86–87, italics in original). However, both change and endurance are dynamic, because endurance is 'the process of continuously inheriting a certain identity of character transmitted throughout a historical route of events' (Whitehead 1925, 108). Environmental contingencies help to determine endurance: 'a favourable environment is essential to the maintenance of a physical object' (Whitehead 1925, 109). However, in the fullness of time, all geomorphic features are fated to change: 'One all pervasive fact, inherent in the very character of what is real is the transition of things, the passage one to another' (Whitehead 1925, 93). In other words, everything in geomorphology (and physical geography for that matter) takes part in an overall process of becoming, being, and fading away. From this process perspective, timeless and timebound views fuse in the dual notions of change and endurance.

Another attempt to reconcile the timeless and timebound schools in geomorphology comes from recent developments in **systems** theory (see **complexity**). Early discussions of nonlinear dynamical systems intimated that some geomorphic systems contain deterministic elements and probabilistic elements. Deterministic elements derive from the universal and necessary operation of geomorphic laws, which apply in all landscapes at all times, though owing to **thresholds**, they may not operate in all landscapes at all times; probabilistic elements derive from historical happenstance and contingency (e.g. Huggett, 1988). The latest expansion of this idea suggests that because of their immanent deterministic–contingent duality, the study of nonlinear geomorphic systems may help to bridge the gap between process and historical studies (Phillips 2006a, 2006b, 2007). The argument is that geomorphic systems have multifarious environmental controls and forcings, which acting in concert can produce many different landscapes. What is more, some controls and forcings are causally contingent and specific to different times and places. Dynamical instability creates and enhances some of this contingency by encouraging the effects of small initial variations and local disturbances to persist and grow disproportionately large, as established in Adrian E. Scheidegger's (1983) instability principle. Now, the combined probability of any particular set of global controls is low, and the probability of any set of local, contingent controls is even lower. As a result, the likelihood of any landscape or geomorphic system existing at a particular place and time is negligibly small – all landscapes are perfect, in the sense that they are an improbable coincidence of several different forces or factors (Phillips 2007). This fascinating notion, which has much in common with evolutionary geomorphology, dispenses with the view that all landscapes and landforms are the inevitable outcome of deterministic laws. Instead, it offers a powerful and integrative new view that sees landscapes and landforms as circumstantial and contingent outcomes of deterministic laws operating in a specific environmental and historical context, with several outcomes possible for each set of processes and boundary conditions.

TOPOGRAPHY

Topography is the lie of the land, or the general configuration of the land surface, including its relief and the location of its features, natural and human-made. It is also the lie of the sea floor and is applicable to the description of submarine relief features. Several locational and structural topographic factors are important in biogeography and ecology, climatology and meteorology, geomorphology, hydrology, and pedology. The influences of latitude, longitude, altitude, and terrain structure are particularly potent. Latitude, longitude, and altitude are locational topographic factors that have basic, if indirect, influences on environmental **systems**. A place's location indirectly determines or constrains the climate it experiences and, to some degree, other environmental conditions. For all these reasons, topography is an immensely important concept in physical geography.

Topography can act directly upon climate, water flow and storage, **soils** and sediments, and living things. Several structural topographic attributes influence meteorological elements and climate. Essentially, topography has a three-dimensional character that alters airflow patterns, so modifying precipitation, cloud, and so on; and that affects the surface radiation balance, modifying the thermal characteristic of the near-surface air and soil. On a large scale, the size and orientation of such features as mountains and plateaux are powerful factors, as is the disposition of land and sea. On a regional scale, relative relief and terrain shape tend to be particularly influential properties.

The three-dimensional form of the land surface on a regional scale affects climates and life. Some mountains develop their own weather systems and characteristic climates. All topographic features disrupt airflow patterns and may generate distinctive atmospheric circulation patterns. Some effects are global, as when north–south trending mountain ranges interfere with planetary waves and instigate the genesis of cyclones. Regional and local effects arise under clear skies and calm conditions, when valleys tend to develop distinctive circulations of air. The regional aspect (orientation) of large topographic features determines exposure to prevailing winds. Leeward slopes, especially on large hills and mountains, normally lie within a rain shadow. Rain shadow effects on vegetation are pronounced in the Basin and Range Province of the United States: the climates of the Great Basin and mountains are influenced by the Sierra Nevada, and the climates of the prairies and plains are semiarid owing to the presence of the Rocky Mountains. In the Cascades, the eastern, leeward slopes are drier than the western, windward slopes. Consequently, the vegetation changes from western and mountain hemlocks (*Tsuga heterophylla* and *Tsuga mertensiana*) and lovely and alpine firs (*Abies amabilis* and *Abies lasiocarpa*) to western larch (*Larix occidentalis*) and ponderosa pine (*Pinus ponderosa*), and finally to sagebrush (*Artemisia tridentata*) desert (Billings 1990). Animals also appear to respond to some regional topographic properties. In western North America, mountain goats (*Oreamnos americanus*) and peregrine falcons (*Falco peregrinus*) prefer to live in steep, rugged terrain, whereas pronghorn antelopes (*Antilocapra americana*) and the greater prairie-chickens (*Tympanuchus cupido*) are confined to flat and gently rolling terrain (Beasom *et al.* 1983).

Local topographic factors influence climate. The chief local factors are aspect and slope of the ground surface; the orientation of such barriers as hedgerows, walls, and buildings; the vertical structure of vegetation and human-made features; the presence of rivers and lakes; and the distribution of vegetation and land-use types. The **microclimates** and **local climates** resulting from such topographic influences in turn affect local water balances, soil **evolution**, geomorphic processes, and the distribution of animals and plants.

Slope aspect produces differences in microclimates and local climates large enough to have a considerable impact on **ecosystems**. A study of forests in Little Laurel Run watershed, West Virginia, USA, found that the north and east aspects were up to 50 per cent more productive than the west and southwest aspects. Moreover, some species groups displayed strong preferences for certain aspects, included yellow poplar (*Liriodendron tulipifera*) and black cherry (*Prunus serotina*) for north and east aspects, and chestnut oak (*Quercus prinus*) and white oak (*Quercus alba*) for west and south-west aspects; however, red oak (*Quercus rubra*) and red maple (*Acer rubrum*), which have a broad **tolerance range,** showed only mild aspect preference (Desta *et al.* 2004). On the south-facing slopes of the Kullaberg Peninsula, south-west Sweden, several arthropod species live far to the north of their main range. The species include the silky wave moth (*Idaea dilutaria*), the beetle *Danacea pallipes*, and the spider *Theridion conigerum* (Ryrholm 1988). Studies of Alpine marmots (*Marmota marmota*) in the French Alps revealed that aspect plays a significant role in **habitat** preference and growth rates (Allainé *et al.* 1994, 1998).

Slope gradient and slope curvature also exert a considerable influence on ecosystems. The sequence of slopes from hilltop to valley bottom produces soil and vegetation **catenas** or toposequences, examples of which are legion. More subtle effects relate to the three-dimensional character of the land surface that dictates divergence of moving water and sediment over noses and convergence in hollows. In the central Appalachian Mountains, pine forest tends to grow on convex nose-slopes, oak forest on side slopes, and northern hardwood species (beech, maple, and birch) in concave hollows. In central Massachusetts, United States, concave slopes have thin loam soils supporting white ash (*Fraxinus americana*), while convex slopes have a thick loam soil supporting the northern red oak (*Quercus rubra*) or the grey oak (*Quercus borealis*) (Hack and Goodlett 1960).

TRANSPORT PROCESSES

The circulation of materials within and between the Earth's spheres requires the transport of substances. The transport of all materials, from solid particles to dissolved ions, needs a force to start and maintain motion. Such forces make boulders fall from cliffs, **soil** and sediment move down hillslopes, water and ice flow along channels, and water move through soil and plants. For this reason, the mechanical principles controlling movement underpin the understanding of transport processes.

The forces that drive sediment movement largely derive from gravity, from climatic effects (heating and cooling, freezing and thawing, winds), and from the action of animals and plants. They may act directly, as in the case of gravity, or indirectly through such agencies as water and wind. In the first case, the force makes the sediment move, as in landslides; in the second case, the force makes the agency move (water for instance) and in turn the moving agency exerts a force on the sediment and tends to move it, as in sediment transport in rivers. The chief forces that act upon geomorphic materials are gravitational forces, fluid forces, water pressure forces, expansion forces, global fluid movements, and biological forces, some of which are expressible as geomorphic transport laws (Table 9). In the atmosphere, the driving forces of motion are pressure gradients, with temperature gradients leading to **convection**. Basically, air moves from areas of excess mass (high pressure) to areas of mass deficit (low pressure), and once in motion the Earth's rotation affects it, making the wind seem to change direction to an observer on the ground (Coriolis force), the balance between the pressure gradient and Coriolis forces giving the geostrophic wind.

Table 9 Geomorphic transport laws

Process	Geomorphic transport law	Mechanisms	Definition of terms
Soil production rate	$P = P_0^{-\alpha d}$	Salt and freeze–thaw weathering, atmospheric dust input, mineral alteration leading to loss of physical strength; bioturbation (animal burrowing, root growth, and tree throw), geochemical reactions mediated by microbes	P, soil production rate from bedrock; d, soil thickness; α, constant
Slope-dependent downslope movement (creep)	$\mathbf{q}_s = K\nabla h$	Wetting and drying, freezing and thawing, shear flow, bioturbation	$\mathbf{q}_s$, volumetric sediment transport rate per unit width; K, constant; h, elevation
Landsliding	None available, but important starts for earthflows, deep-seated landslides, and landslide dynamics	Stress exceeds material strength owing to earthquakes, elevated pore pressures derived from precipitation or from undermining of toe; released sediment travels downslope	–

continued overleaf

Table 9 Continued

Process	Geomorphic transport law	Mechanisms	Definition of terms
Surface wash and splash	Many short-term empirical and mechanistic expressions, but no geomorphic transport law available	Rainsplash and overland flow displace and remove particles; rill and gully incision	–
River incision into bedrock	$E = k_b\, A^m S^n$	Plucking and particle wear due to river flow and sediment transport	E, incision rate; k_b, constant that may depend on uplift rate and rock strength; A, drainage area; S, local slope; m, n, constants
Debris flow incision into bedrock	$E = k_d f\left[\rho_s D^2 \left(\dfrac{\partial u}{\partial y}\right)^a L_s\right]^p$	Particle impact and sliding wear of bedrock during mass transport	D, representative grain diameter; k_d, constant that depends on bedrock properties; f, frequency of debris flows; ρ_s, bulk density of debris flow; u, debris flow velocity as a function of distance y above the bed; L_s, length of debris flow 'snout'; a, p, constants
Glacial scour	$E = c U_b$	Sediment-rich basal sliding wears bedrock	U_b, basal ice velocity; and c, constant
Wind transport and scour	Extensive theory for sediment transport by wind; some theory for rock abrasion	Abrasion by wind-suspended particles	–

Source: Adapted from Dietrich and Perron (2006)

In oceans and seas, various forces acting upon the water (planetary rotation, winds, temperature gradients, salinity gradients, the gravitational pull of the Sun and Moon) set up currents (see **general circulation of the oceans**). In soil, water drains through larger pores and channels owing to gravitational forces and moves through the smaller soil matrix owing to water potentials. Soil water potentials have two components – the matrix potential (suction), which depends upon the physical character of the soil particles, and the osmotic potential (suction), which depends on the presence of solutes. Within the plumbing system of plants, water moves mainly because of the potential difference between the evaporation of water from the leaves (transpiration) and the water potential at the root surfaces.

UNIFORMITARIANISM

Many see uniformitarianism as the foundation stone of modern physical geography. This is because, in practising their trade, scientists follow rules or guidelines established by scientific practitioners. These guidelines advise scientists how to go about the business of making scientific inquiries. In other words, they are guidelines concerned with scientific methodology or procedures. The first rule is the uniformity of law, which almost all scientists unquestioningly follow. It is the supposition that natural laws are invariant in time and space; or, to put it another way, it is the postulate that the properties of **energy** and matter have been the same in the past as they are at present. The uniformity of law is the most indispensable part of the scientist's creed. If laws are changeable, then science cannot proceed for determinism fails. Charles Lyell, the nineteenth-century geologist, persuasively advocated three 'uniformities', as well as the uniformity of law: the uniformity of process (**actualism**), the uniformity of rate (**gradualism**), and the uniformity of state (steady-statism). These four rules constitute Lyell's uniformitarianism (see Gould 1977).

A common blunder is to equate uniformitarianism with actualism. Uniformitarianism was a system of assumptions about Earth history, one of which was the assumption of actualism (the uniformity of process): Lyell's uniformitarian dogma was a set of beliefs about Earth surface processes and states that involved more than actualism. Lyell postulated that present processes have occurred throughout Earth history, have run at a gradual rate, and have not caused the Earth to change in any particular direction. Other sets of beliefs are possible. The diametric opposite of Lyell's uniformitarian position would be a belief in the non-uniformity of process (non-actualism), the non-uniformity of rate (**catastrophism**), and the non-uniformity of state (**directionalism**). All other combinations of assumption are possible and give rise to different 'systems of Earth history' (Huggett 1997b). The various systems are testable by field evidence. To be sure, directionalism was accepted even before Lyell's death, and non-actualism and, in particular, catastrophism are discussed in Earth and biological sciences.

Further reading: Gould 1965; Huggett 1997b

VICARIANCE

Vicariance is simply the splitting of **populations** or taxonomic groups (species, genera, families, and so forth) into two or more isolated parts. It may occur across the full range of geographical scales, from a motorway dividing a population of frogs into two separate subpopulations, to the breaking up of supercontinents that sunders entire faunas and floras, which subsequently occupy isolated landmasses. Although it is at root a simple process, its invocation to explain large-scale biogeographical patterns has caused a fiery debate with centre-of-origin–**dispersal** biogeographers. Ever since Charles Darwin and Alfred Russel Wallace, dispersal biogeographers such as William Diller Matthew, George Gaylord Simpson, and Philip J. Darlington have sought to explain the global distribution of species by assuming that each species arises in a specific place (its centre of origin) and then disperses elsewhere by crossing such obstacles as oceans and mountain ranges (see **dispersal**). Léon Croizat, a Franco-Italian scholar, directed the first major challenge to this ruling biogeographical theory (Croizat 1958, 1964; see also Humphries 2000). He tested the centre-of-origin–dispersal model by mapping the distributions of hundreds of plant and animal species, and found that species with quite different dispersal propensities and colonizing abilities had the same pattern of geographical distribution. He termed these shared geographical distributions generalized or standard tracks, and reasoned that, rather than representing lines of migration, they are the present distributions of a set of ancestral distributions, or a biota of which individual components are relict fragments. His reasoning was that climatic, sea level, and tectonic changes had fragmented widespread ancestral taxa. The fragmentation process eventually acquired the name vicariance. Of course, to become widespread in the first place, a species must disperse, but vicariance biogeographers claim that ancestral taxa achieve widespread distribution through a mobile phase *in the absence* *of barriers*. They allow that some dispersal across barriers does occur, but feel that it is a relatively insignificant biogeographical process. So, the key difference between dispersal and vicariance explanations lies in the nature of barriers to spread: a vicariance event splits the ancestral range of a taxon by creating a barrier, while dispersal extends an ancestral range over a pre-existing barrier.

To grasp the idea of vicariance, consider a taxon with a disjunct distribution. The genus *Nothofagus* (the southern beeches) consists of about 60 species of evergreen and deciduous trees and shrubs, and its present distribution is disjunct, being found on remnants of Gondwana – South America, New Zealand, Australia, New Caledonia, and New Guinea, and Antarctica (fossil pollen of Oligocene age), but not Africa. As this plant has seeds unsuited to long-distance dispersal, the conclusion normally drawn is that the modern disjunct range of the genus has resulted from the breakup of Gondwana, which was a vicariance event. Similarly, New World geckos appear to have split from their Old World relatives around 96 million years ago when South America broke away from Africa as the Atlantic Ocean opened (Gamble *et al.* 2008). On the other hand, tapirs (close relatives of horses and rhinoceroses) have a disjunct distribution with three species in central and South America and one species in South-East Asia. In this case, the presence of the oldest tapir fossils in Europe suggests that tapirs originated there, dispersed to the Americas and to South-East Asia, and then became extinct save in their present locations.

Evidence is emerging in Europe and North America of vicariance caused by the position of ice sheets during last glacial maximum some 18,000 years ago. At that time in Europe, an ice sheet covered Scandinavia and Finland while tundra occurred over most of the rest of northern Eurasia. Populations of boreal plants survived in more or less disjunct **refugia** in southern Europe, which often lay between mountain ranges and seas. Evidence from genetics and palaeobotany indicates that these populations, which included Scots pine (*Pinus sylvestris*), contributed a good deal to the recolonization of more northern latitudes during the Holocene. Fractures in the present genetic structure of Scots pine suggest such glacial vicariance (Naydenov *et al.* 2007). Likewise, the Canadian side of the Pacific Northwest of North America was almost entirely ice-covered during the last glacial maximum, and this appears to have caused vicariance as populations of several plant and animal taxa, including lodgepole pine (*Pinus contorta*), survived in ice-free refugia (Godbout *et al.* 2008).

Despite the undoubted success of the vicariance model in explaining the biogeographical history of some groups of organisms,

dispersal, and especially oceanic dispersal, is making a strong comeback as a key process in biogeography.

Further reading: Nelson and Rosen 1981

ZONALITY

The idea of zones and zonality finds many applications in physical geography and is applicable at different geographical scales. On a global scale, there are climatic zones, vegetation zones, **soil** zones, and so forth. Planetary fields of temperature and wind create three basic climatic zones: the torrid zone of low latitudes, the temperate zones of middle latitudes, and the frigid zones of high latitudes. The distribution of land and sea and the presence of large-scale topographic features, chiefly mountain ranges and plateaux, distort this fundamental zonary arrangement of climates. At present, the distortions to the three basic climatic zones produce some seven climatic regions, though the exact number varies from one authority to the next. The regions are: humid tropical climate (equatorial rain zone); savannah climate (tropical margin zone with summer rains); desert climate (subtropical dry zone); Mediterranean climate (subtropical zone of winter rain and summer drought); temperate climate (temperate zone with precipitation all year round), which is divisible into maritime climate (warm temperate subzone), nemoral climate (typical temperate subzone – a short period of frost), continental climate (arid temperate subzone – cold winter), and boreal climate (cold temperate subzone); tundra climate (subpolar zone); and polar climate (polar zone). These climatic zones largely determine the distribution of zonal soil groups and zonal vegetation communities (zonobiomes), weathering zones, and, to an questionable extent, some types of landform.

Altitudinal zones mirror latitudinal zones. As the general decline in temperature with increasing latitude produces latitudinal climatic zones, so decreasing temperature with increasing elevation produces altitudinal climatic zones. Starting at a mountain base, the zones are usually submontane, montane, subalpine, alpine, and nival. In the tropics, altitudinal effects are strong enough to allow polar conditions on mountaintops, with permanent snow in, for example, the equatorial Andes and New Guinea and African mountains, and tropical glaciers on, for instance, Mount Kilimanjaro. Moreover, the altitudinal depression of temperature defines the character of many significant tropical places that are around or over 2,000 m above sea level, such as Mexico City, Johannesburg, Nairobi, Quito, and La Paz.

Altitudinal climatic zones, like their latitudinal counterparts, influence soils, landforms, and vegetation.

Zones at more local scales occur in various environmental **systems**. Many features of landforms, soils, and vegetation have zonal characteristics, as in zones on a beach and along hillslopes. Moving landwards, the nearshore zone of a coast comprises breaker, surf, and swash zones. In the arctic–alpine **environment** of Greenland, a typical geomorphic toposequence comprises four geomorphodynamic zones, each associated with a particular prevailing process regime (Stablein 1984). First is the zone of peaks, plateaux, and saddles where frost weathering, solution weathering, slope wash, cryoturbation, and wind deflation are the chief processes and create such landforms as weathering pans, solution hollows, and tafoni (hemispherical hollows in rock surfaces), and, in more sheltered sites, where enough frost-weathered debris and moisture accumulate, patterned ground. The second zone is the upper slope zone, where frost weathering, cryogenic rockfall, and debris creep are the main processes. Slope wash, solifluction, nivation, and slope dissection characterize the third, or midslope, zone where landforms include gelifluction lobes, solifluction lobes, and sorted stone stripes. At the foot of the sequence is the lower slope zone where cryoturbation, frost heaving, and wind deflation dominate, and gelifluction and solifluction feed valley-side sediment into the valley bottom; landforms found in this zone include terraces and bare rock outcrops.

Further reading: Huggett and Cheesman 2002

BIBLIOGRAPHY

Aandahl, A. R. (1948) The characterization of slope positions and their influence on the total nitrogen content of a few virgin soils of western Iowa. *Soil Science Society of America Proceedings* 13, 449–54.

Akbari, H. (2009) *Urban Heat Islands*. Berlin and Heidelberg: Springer-Verlag.

Albritton, C. C., Jr (1989) *Catastrophic Episodes in Earth History*. London and New York: Chapman & Hall.

Allainé, D., Rodrigue, I., Le Berre, M., and Ramousse, R. (1994) Habitat preferences of alpine marmots, *Marmota marmota*. *Canadian Journal of Zoology* 72, 2193–98.

Allainé, D., Graziani, L., and Coulon, J. (1998) Postweaning mass gain in juvenile alpine marmots *Marmota marmota*. *Oecologia* 113, 370–76.

Amundson, R. and Jenny, H. (1991) The place of humans in the state factor theory of ecosystems and their soils. *Soil Science* 151, 99–109.

Anderson, D. L. (2005) Scoring hotspots: the plume and plate paradigms. In G. R. Foulger, J. H. Natland, D. C. Presnall, and D. L. Anderson (eds) *Plates, Plumes, and Paradigms* (Geological Society of America Special Paper 388), pp. 31–54. Boulder, CO: The Geological Society of America.

Arnold, E. N. (1994) Investigating the origins of performance advantage: adaptation, exaptation and lineage effects. In P. Eggleton and R. I. Vane-Wright (eds) *Phylogenetics and Ecology*, pp. 124–68. London: Academic Press.

Ashman, M. R. and Puri, G. (2002) *Essential Soil Science: A Clear and Concise Introduction to Soil Science*. Oxford: Blackwell.

Aubréville, A. (1949) *Climats, Forêts et Désertification de l'Afrique Tropicale*. Paris: Société d'Éditions Géographiques, Maritimes et Coloniales.

Baas, A. C. W. (2002) Chaos, fractals and self-organization in coastal geomorphology: simulating dune landscapes in vegetated environments. *Geomorphology* 48, 309–28.

Bailey, R. G. (1995) *Description of the Ecoregions of the United States*, 2nd edn, revised and enlarged (Miscellaneous Publication No. 1391). Washington, DC: United States Department of Agriculture, Forest Service.

—— (1996) *Ecosystem Geography*, With a foreword by Jack Ward Thomas, Chief, USDA Forest Service. New York: Springer.

—— (1997) *Ecoregions: The Ecosystem Geography of Oceans and Continents*. New York: Springer-Verlag.

—— (2002) *Ecoregion-based Design for Sustainability*. New York: Springer-Verlag.

Barrell, J. (1917) Rhythms and the measurement of geologic time. *Bulletin of the Geological Society of America* 28, 745–904.

Barrett, G. W., Van Dyne, G. M., and Odum, E. P. (1976) Stress ecology. *Bio-Science* 26, 192–94.

Barrow, C. J. (1991) *Land Degradation: Development and Breakdown of Terrestrial Environments*. Cambridge: Cambridge University Press.

Barry, R. G. and Chorley, R. J. (2003) *Atmosphere, Weather and Climate*, 8th edn. London: Routledge.

Baumgartner, A. and Reichel, E. (1975) *The World Water Balance*. Amsterdam and Oxford: Elsevier.

Beard, J. S. (2003) Paleodrainage and the geomorphic evolution of passive margins in southwestern Australia. *Zeitschrift für Geomorphologie* NF 47, 273–88.

Beasom, S. L., Wiggers, E. P., and Giardino, J. R. (1983) A technique for assessing land surface ruggedness. *Journal Wildlife Management* 47, 1163–66.

Belton, M. J. S., Morgan, T. H., Samarasinha, N. H., and Yeomans, D. K. (eds) (2004) *Mitigation of Hazardous Comets and Asteroids*. Cambridge: Cambridge University Press.

Benestad, R. E. (2002) *Solar Activity and Earth's Climate*. London, Berlin, Heidelberg, New York: Springer.

Benton, M. J. (2003) *When Life Nearly Died: The Greatest Mass Extinction of All Time*. London: Thames and Hudson.

Berry, P. M., Dawson, T. P., Harrison, P. A., and Pearson, R. G. (2002) Modelling potential impacts of climate change on the bioclimatic envelope of species in Britain and Ireland. *Global Ecology & Biogeography* 11, 453–62.

Berry, P. M., Dawson, T. P., Harrison, P. A., Pearson, R., and Butt, N. (2003) The sensitivity and vulnerability of terrestrial habitats and species in Britain and Ireland to climate change. *Journal for Nature Conservation* 11, 15–23.

Billings, W. D. (1990) The mountain forests of North America and their environments. In C. B. Osmond, L. F. Pitelka, and G. M. Hidy (eds) *Plant Biology of the Basin and Range* (Ecological Studies, vol. 80), pp. 47–86. Berlin: Springer-Verlag.

Birkeland, P. W. (1990) Soil–geomorphic research – a selective review. *Geomorphology* 3, 207–24.

Błazejczyk, K. and Grzybowski, J. (1993) Climatic significance of small aquatic surfaces and characteristics of the local climate of Suwałki Landscape Park (north-east Poland). *Ekologia Polska* 41, 105–21.

Blume, H.-P. and Schlichting, E. (1965) The relationships between historical and experimental pedology. In E. G. Hallsworth and D. V. Crawford (eds) *Experimental Pedology*, pp. 340–53. London: Butterworths.

Bobrowsky, P. T. and Rickman, H. (eds) (2006) *Comet/Asteroid Impacts and Human Society: An Interdisciplinary Approach*. Berlin and Heidelberg: Springer-Verlag.

Bogaert, J., Salvador-Van Eysenrode, D., Impens, I., and Van Hecke, P. (2001a) The interior-to-edge breakpoint distance as a guideline for nature conservation policy. *Environmental Management* 27, 493–500.

Bogaert, J., Salvador-Van Eysenrode, D., Van Hecke, P., Impens, I. (2001b) Geometrical considerations for evaluation of reserve design. *Web Ecology* 2, 65–70.

Bowen, D. Q. (1973) The Quaternary deposits of the Gower. *Proceedings of the Geologists' Association* 84, 249–72.

Bracken, L. J. and Wainwright, J. (2006) Geomorphological equilibrium: myth and metaphor? *Transactions of the Institute of British Geographers*, New Series 31, 167–78.

Bradbury, R. H., Van Der Laan, J. D., and Green, D. G. (1996) The idea of complexity in ecology. *Senckenbergiana Maritima* 27, 89–96.

Brady, N. C. and Weil, R. R. (2007) *The Nature and Properties of Soils*, 14th edn. Upper Saddle River, NJ: Prentice Hall.

Braun, H., Christl, M., Rahmstorf, S., Ganopolski, A., Mangini, A., Kubatzki, C., Roth, K., and Kromer, B. (2005) Possible solar origin of the 1,470-year glacial climate cycle demonstrated in a coupled model. *Nature* 438, 208–11.

Bridges, E. M., Hanman, I. D., Oldeman, L. R., Penning de Vries, F. W. T., Scherr, S. J., and Sombatpanit, S. (eds) (2001) *Response to Land Degradation*. Enfield, NH: Science Publishers.

Broecker, W. S. (1965) Isotope geochemistry and the Pleistocene climatic record. In H. E. Wright Jr and D. G. Frey (eds) *The Quaternary of the United States*, pp. 737–53. Princeton, NJ: Princeton University Press.

—— (1995) Chaotic climate. *Scientific American* 273, 44–50.

Broecker, W. S. and Denton, G. H. (1990) What drives glacial cycles? *Scientific American* 262, 42–50.

Broecker, W. S., Thurber, D. L., Goddard, J., Ku, T., Matthews, R. K., and Mesolella, K. J. (1968) Milankovitch hypothesis supported by precise dating of coral reefs and deep-sea sediments. *Science* 159, 1–4.

Brown, D. J (2006) A historical perspective on soil–landscape modeling. In S. Grunwald (ed.) *Soil–landscape Modeling: Geographical Information Technologies and Pedometrics*, pp. 61–104. New York: CRC Press.

Brown, J. H. (1971) Mammals on mountaintops: nonequilibrium insular biogeography. *The American Naturalist* 105, 467–78.

Brown, J. H. and Lomolino, M V. (1998) *Biogeography*, 2nd edn. Sunderland, MA: Sinauer Associates.

Brown, J. H., Riddle, B. R., and Lomolino, M. V. (2005) *Biogeography*, 3rd edn. Sunderland, MA: Sinauer Associates.

Brunckhorst, D. (2000) *Bioregional Planning: Resource Management Beyond the New Millennium*. Sydney, Australia: Harwood Academic Publishers.

Brunsden, D. (2001) A critical assessment of the sensitivity concept in geomorphology. *Catena* 42, 99–123.

Brunsden, D. and Kesel, R. H. (1973) The evolution of the Mississippi River bluff in historic time. *Journal of Geology* 81, 576–97.

Büdel, J. (1957) Die 'Doppelten Einebnungsflächen' in den feuchten Tropen. *Zeitschrift für Geomorphologie* NF 1, 201–28.

—— (1982) *Climatic Geomorphology*. Translated by Lenore Fischer and Detlef Busche. Princeton, NJ: Princeton University Press.

Buol, S. W., Southard, R. J., Graham, R. C., and McDaniel, P. A. (2003) *Soil Genesis and Classification*, 5th edn. Ames, IA: Iowa State Press.

Burroughs, W. J. (2007) *Climate Change: A Multidisciplinary Approach*, 2nd edn. Cambridge: Cambridge University Press.

Burt, T. and Goudie, A. (1994) Timing shape and shaping time. *Geography Review* 8, 25–29.

Butler, D. R. (1992) The grizzly bear as an erosional agent in mountainous terrain. *Zeitschrift für Geomorphologie* NF 36, 179–89.

—— (1995) *Zoogeomorphology: Animals as Geomorphic Agents*. Cambridge: Cambridge University Press.

Calvert, A. M., Amirault, D. L., Shaffer, F., Elliot, R., Hanson, A., McKnight, J., and Taylor P. D. (2006) Population assessment of an endangered shorebird: the Piping Plover (*Charadrius melodus melodus*) in eastern Canada. *Avian Conservation and Ecology – Écologie et conservation des oiseaux* 1(3), Article 4. Online: www.ace-eco.org/vol1/iss3/art4

Campbell, D. E. (1998) Energy analysis of human carrying capacity and regional sustainability: an example using the State of Maine. *Environmental Monitoring and Assessment* 51, 531–69.

Carson, R. (1962) *Silent Spring*. Boston, MA: Houghton Mifflin.

Censky, E. J., Hodge, K., and Dudley, J. (1998) Over-water dispersal of lizards due to hurricanes. *Nature* 395, 556.

Chapman, C. R. (1996) Book review of *Rogue Asteroids and Doomsday Comets* by D. Steel. *Meteoritics and Planetary Science* 31, 313–14.

—— (2004) The hazard of near-Earth asteroid impacts on earth. *Earth and Planetary Science Letters* 222, 1–15.

Chase, J. M. and Leibold, M. A. (2003) *Ecological Niches: Linking Classical and Contemporary Approaches*. Chicago, IL and London: University of Chicago Press.

Chen, Z.-S., Hsieh, C.-Fu., Jiang, F.-Y., Hsieh, T.-H., Sun, I-F. (1997) Relations of soil properties to topography and vegetation in a subtropical rain forest in southern Taiwan. *Plant Ecology* 132, 229–41.

Chorley, R. J. (1962) *Geomorphology and General Systems Theory* (US Geological Survey Professional Paper 500-B). Washington, DC: United States Government Printing Office.

—— (1965) A re-evaluation of the geomorphic system of W. M. Davis. In R. J. Chorley and P. Haggett (eds) *Frontiers in Geographical Teaching*, pp. 21–38. London: Methuen.

—— (1969) The drainage basin as the fundamental geomorphic unit. In R. J. Chorley (ed.) *Water, Earth, and Man: A Synthesis of Hydrology, Geomorphology, and Socio-economic Geography*, pp. 77–99. London: Methuen.

Chorley, R. J., and Kennedy, B. A. (1971) *Physical Geography: A Systems Approach*. London: Prentice-Hall.

Chorley, R. J., Beckinsale, R. P., and Dunn A. J. (1973) *The History of the Study of Landforms: Volume 2, The Life and Work of William Morris Davis*. London: Methuen.

Church, M. and Mark, D. M. (1980) On size and scale in geomorphology. *Progress in Physical Geography* 4, 342–90.

Clausen, J. (1965) Population studies of alpine and subalpine races of conifers and willows in the California high Sierra Nevada. *Evolution* 19, 56–68.

Clements, F. E. (1916) *Plant Succession: An Analysis of the Development of Vegetation* (Carnegie Institute of Washington, Publication No. 242). Washington, DC: Carnegie Institute of Washington.

Cocks, L. R. M. and Parker, A. (1981) The evolution of sedimentary environments. In L. R. M. Cocks (ed.) *The Evolving Earth*, pp. 47–62. Cambridge: Cambridge University Press; London: British Museum (Natural History).

Coe, A. L., Bosence, D. W. J., Church, K. D., Flint, S. S., Howell, J. A., and Wilson, R. C. L. (2003) *The Sedimentary Record of Sea-Level Change*. Cambridge: Cambridge University Press.

Cohen, J. E. (1996) *How Many People Can the Earth Support?* New York and London: W. W. Norton.

Colinvaux, P. A., De Oliveira, P. E., Moreno, J. E., Miller, M. C., and Bush, M. B. (1996) A long pollen record from lowland Amazonia: forest and cooling in glacial times. *Science* 274, 85–88.

Colman, S. M. and Pierce, K. L. (2000) Classification of Quaternary geochronologic methods. In J. S. Noller, J. M. Sowers, and W. R. Lettis (eds) *Quaternary Geochronology: Methods and Applications* (AGU Reference Shelf 4), 2–5. Washington, DC: American Geophysical Union.

Coope, G. R. (1994) The response of insect faunas to glacial–interglacial climatic fluctuations. *Philosophical Transactions of the Royal Society of London* 344B, 19–26.

Cooper, W. S. (1923) The recent ecological history of Glacier Bay, Alaska. *Ecology* 6, 197.

Corstanje, R., Grunwald, S., Reddy, K. R., Osborne, T. Z., and Newman, S. (2006) Assessment of the spatial distribution of soil properties in a northern Everglades marsh. *Journal of Environmental Quality* 35, 938–49.

Cowie, J. (2007) *Climate Change: Biological and Human Aspects.* Cambridge: Cambridge University Press.

Cowie, R. H. and Holland, B. S. (2006) Dispersal is fundamental to biogeography and the evolution of biodiversity on oceanic islands. *Journal of Biogeography* 33, 193–98.

Cowles, H. C. (1899) The ecological relations of the vegetation on the sand dunes of Lake Michigan. *Botanical Gazette* 27: 95–117, 167–202, 281–308, 361–91.

Cox, G. W. (1999) *Alien Species in North America and Hawaii: Impacts on Natural Ecosystems.* Washington, DC: Island Press.

Coxson, D. S. and Marsh, J. (2001) Lichen chronosequences (postfire and post-harvest) in lodgepole pine (*Pinus contorta*) forests of northern interior British Columbia. *Canadian Journal of Botany* 79, 1449–64.

Coyne, J. A. and Orr, H. A. (2004) *Speciation.* Sunderland, MA: Sinauer Associates.

Creemans, D. L., Brown, R. B., and Huddleston, J. H. (eds) (1994) *Whole Regolith Pedology* (Soil Science Society of America Special Publication 34). Madison, WI: Soil Science Society of America.

Crocker, R. L. and Major, J. (1955) Soil development in relation to vegetation and surface age at Glacier Bay, Alaska. *Journal of Ecology* 43, 427–48.

Croizat, L. (1958) *Pangeography*, 2 vols. Caracas: Published by the author.

—— (1964) *Space, Time, Form: The Biological Synthesis.* Caracas: Published by the author.

Croteau, M.-N., Luoma, S. N., and Stewart, A. R. (2005) Trophic transfer of metals along freshwater food webs: evidence of cadmium biomagnification in nature. *Limnology and Oceanography* 50, 1511–19.

Culling, W. E. H. (1987) Equifinality: modern approaches to dynamical systems and their potential for geographical thought. *Transactions of the Institute of British Geographers*, New Series 12, 57–72.

—— (1988) A new view of the landscape. *Transactions of the Institute of British Geographers*, New Series 13, 345–60.

Currie, R. G. (1984) Evidence for 18.6-year lunar nodal drought in western North America during the past millennium. *Journal of Geophysical Research* 89, 1295–308.

Dale, V. H., Joyce, L. A., McNulty, S., Neilson, R. P., Ayres, M. P., Flannigan,

M. D., Hanson, P. J., Irland, L. C., Lugo, A. E., Peterson, C. J., Simberloff, D., Swanson, F. J., Stocks, B. J., and Wotton, B. M. (2001) Climate change and forest disturbances. *BioScience* 51, 723–34.

Dale, V. H., Swanson, F. J., and Crisafulli, C. M. (2005) *Ecological Responses to the 1980 Eruption of Mount St. Helens*. New York: Springer.

Darwin, C. R. (1859) *The Origin of Species by Means of Natural Selection, or the Preservation of Favoured Races in the Struggle for Life*. London: John Murray.

—— (1881) *The Formation of Vegetable Mould through the Action of Worms, with Observations on Their Habits*. London: John Murray.

Davies, G. F. (1999) *Dynamic Earth: Plates, Plumes and Mantle Convection*. Cambridge: Cambridge University Press.

Davis, W. M. (1899) The geographical cycle. *Geographical Journal* 14, 481–504. (Also in *Geographical Essays*)

—— (1909) *Geographical Essays*. Boston, MA: Ginn.

Décamps, H. (2001) How a riparian landscape finds form and comes alive. *Landscape and Urban Planning* 57, 169–75.

Delcourt, H. R. and Delcourt, P. A. (1988) Quaternary landscape ecology: relevant scales in space and time. *Landscape Ecology* 2, 23–44.

—— (1994) Postglacial rise and decline of *Ostrya virginiana* (Mill.) K. Koch and *Carpinus caroliniana* Walt. in eastern North America: predictable responses of forest species to cyclic changes in seasonality of climate. *Journal of Biogeography* 21, 137–50.

Delong, D. C. Jr (1996) Defining biodiversity. *Wildlife Society Bulletin* 24, 738–49.

Desta, F., Colbert, J. J., Rentch, J. S., and Gottschalk, K. W. (2004) Aspect induced differences in vegetation, soil, and microclimatic characteristics of an Appalachian watershed. *Castanea* 69, 92–108.

Dickinson, G. and Murphy, K. (2007) *Ecosystems*, 2nd edn. London: Routledge.

Dieckmann, U. and Doebeli, M. (1999) On the origin of species by sympatric speciation. *Nature* 400, 354–57.

Dieckmann, U., Doebeli, M., Metz, J. A. J., and Tautz, D. (eds) (2004) *Adaptive Speciation*. Cambridge: Cambridge University Press.

Dietrich, W. E. and Perron, J. T. (2006) The search for a topographic signature of life. *Nature* 439, 411–18.

Dobzhansky, T. (1937) *Genetics and the Origin of Species*. New York: Columbia University Press.

Dott, R. H. (ed.) (1992) *Eustasy: The Historical Ups and Downs of a Major Geological Concept* (Geological Society of America Memoir 180). Boulder, CO: The Geological Society of America.

Douglas, B., Kearney, M. S., and Leatherman, S. P. (eds) (2001) *Sea Level Rise: History and Consequences* (International Geophysics Series, vol. 75). Foreword by John Knauss. San Diego, CA and London: Academic Press.

Drake, J. A. (1990) The mechanics of community assembly and succession. *Journal of Theoretical Biology* 147, 213–33.

Drury, W. H. and Nisbet, I. C. T. (1973) Succession. *Journal of the Arnold Arboretum* 54, 331–68.

Dunn, G. E. (1940) Cyclogenesis in the tropical Atlantic. *Bulletin of the American Meteorological Society* 21, 215–29.

Dury, G. H. (1969) Relation of morphometry to runoff frequency. In R. J. Chorley (ed.) *Water, Earth, and Man: A Synthesis of Hydrology, Geomorphology, and Socio-Economic Geography*, pp. 419–30. London: Methuen.

Eddy, J. A. (1977a) Anomalous solar radiation during the seventeenth century. *Science* 198, 824–29.

—— (1977b) The case of the missing sunspots. *Scientific American* 236, 80–92.

—— (1977c) Climate and the changing Sun. *Climatic Change* 1, 173–90.

—— (1983) The Maunder minimum: a reappraisal. *Solar Physics* 89, 195–207.

Eldredge, N. and Gould, S. J. (1972) Punctuated equilibria: an alternative to phyletic gradualism. In T. J. M. Schopf (ed.), *Models in Paleobiology*, pp. 82–115. San Francisco, CA: Freeman, Cooper.

Eliot, C. (2007) Method and metaphysics in Clements's and Gleason's ecological explanations. *Studies in History and Philosophy of Biological and Biomedical Sciences* 38, 85–109.

Elkibbi, M. and Rial, J. A. (2001) An outsider's review of the astronomical theory of the climate: is the eccentricity-driven insolation the main driver of the ice ages? *Earth-Science Reviews* 56, 161–77.

Elkins-Tanton, L. T. (2005) Continental magmatism caused by lithospheric delamination. In G. R. Foulger, J. H. Natland, D. C. Presnall, and D. L. Anderson (eds) *Plates, Plumes, and Paradigms* (Geological Society of America Special Paper 388), pp. 449–61. Boulder, CO: The Geological Society of America.

Elkins-Tanton, L. T. and Hager, B. H. (2005) Giant meteoroid impacts can cause volcanism. *Earth and Planetary Science Letters* 239, 219–32.

Elton, C. S. (1927) *Animal Ecology*. London: Sidgwick and Jackson.

—— (1958) *The Ecology of Invasions by Animals and Plants*. London: Chapman & Hall.

Emanuel, K. A. (1994) *Atmospheric Convection*. Oxford: Oxford University Press.

Engebretson, D. C., Kelley, K. P., Cashman, H. J., and Richards, M. A. (1992) 180 million years of subduction. *GSA–Today* 2, 93–95, and 100.

Erwin, D. H. (2006) *Extinction: How Life on Earth nearly ended 250 Million Years Ago*. Princeton, NJ: Princeton University Press.

Falkowski, P., Scholes, R. J., Boyle, E., Canadell, J., Canfield, D., Elser, J., Gruber, N., Hibbard, K., Högberg, P., Linder, S., Mackenzie, F. T., Moore III, B., Pedersen, T., Rosenthal, Y., Seitzinger, S., Smetacek, V., and Steffen, W. (2000) The global carbon cycle: a test of our knowledge of Earth as a system. *Science* 290, 291–96.

Fastie, C. L. (1995) Causes and ecosystem consequences of multiple pathways of ecosystem succession at Glacier Bay, Alaska. *Ecology* 76, 1899–1916.

Favis-Mortlock, D. and de Boer, D. (2003) Simple at heart? Landscape as a self-organizing complex system. In S. Trudgill, and A. Roy (eds) *Contemporary Meanings in Physical Geography: From What to Why?*, pp. 127–71. London: Arnold.

Fazey, I., Fischer, J., and Lindenmayer, D. B. (2005) What do conservation biologists publish? *Biological Conservation* 124, 63–73.

Fernández, M. H. and Peláez-Campomanes, P. (2003) The bioclimatic model: a method of palaeoclimatic qualitative inference based on mammal associations. *Global Ecology & Biogeography* 12, 507–17.

Fischer, J. and Lindenmayer, D. B. (2007) Landscape modification and habitat fragmentation: a synthesis. *Global Ecology & Biogeography* 16, 265–80.

Flannery, T. F., Rich, T. H., Turnbull, W. D., and Lundelius, E. L., Jr (1992) The Macropodoidea (Marsupialia) of the early Pliocene Hamilton local fauna,

Victoria, Australia. *Fieldiana: Geology*, New Series No. 25. Chicago, IL: Field Museum of Natural History.

Foukal, P., Fröhlich, C., Spruit, H., and Wigley, T. M. L. (2006) Variations in solar luminosity and their effect on the Earth's climate. *Nature* 443, 161–66.

Foulger, G. R. (2005) Mantle plumes: why the current skepticism? *Chinese Science Bulletin* 50, 1555–60.

Foulger, G. R., Natland, J. H., Presnall, D. C., and Anderson, D. L. (eds) (2005) *Plates, Plumes, and Paradigms* (Geological Society of America Special Paper 388). Boulder, CO: The Geological Society of America.

French, B. M. (1998) *Traces of Catastrophe: A Handbook of Shock-Metamorphic Effects in Terrestrial Meteorite Impact Structures* (LPI Contribution No. 954). Houston, TX: Lunar and Planetary Institute.

Fry, C. (2007) *The Impact of Climate Change: The World's Greatest Challenge in the Twenty-first Century*. London: New Holland Publishers.

Fukao, Y., Maruyama, S., Obayashi, M., and Inoue, H. (1994) Geologic implication of the whole mantle P-wave tomography. *Journal of the Geological Society of Japan* 100, 4–23.

Futuyma, D. J. (2005) *Evolution*. Sunderland, MA: Sinauer Associates.

Gamble, T., Bauer, A. M., Greenbaum, E., and Jackman, T. R. (2008) Evidence for Gondwanan vicariance in an ancient clade of gecko lizards. *Journal of Biogeography* 35, 88–104.

Gartland, L. (2008) *Heat Islands: Understanding and Mitigating Heat in Urban Areas*. London: Earthscan Publications.

Gasperini, L., Bonatti, E., and Longo, G. (2008) The Tunguska mystery: finding a piece of the elusive cosmic body that devastated a Siberian forest a century ago could help save the Earth in the centuries to come. *Scientific American* 298 (June), 80–87.

Gaston, K. J. and Spicer, J. I. (2004) *Biodiversity: An Introduction*, 2nd edn. Oxford: Blackwell Publishing.

Geeson, N. A., Brandt, C. J., and Thornes, J. B. (eds) (2002) *Mediterranean Desertification: A Mosaic of Processes and Responses*. Chichester: John Wiley & Sons.

Geist, H. (2005) *The Causes and Progression of Desertification*. Aldershot, Hampshire: Ashgate Publishing.

Gell-Mann, M. (1994) *The Quark and the Jaguar: Adventures in the Simple and the Complex*. New York: W. H. Freeman.

Gerday, C. and Glansdorff, N. (2007) *Physiology and Biochemistry of Extremophiles*. Washington, DC: AMS Press, American Society for Microbiology.

Gilbert, G. K. (1877) *Geology of the Henry Mountains (Utah)* (United States Geographical and Geological Survey of the Rocky Mountains Region). Washington, DC: United States Government Printing Office.

Gingerich, P. D. (2001) Rates of evolution on the time scale of the evolutionary process. *Genetica* 112/113, 127–44.

Givnish, T. J. and Sytsma, K. J. (eds) (1997) *Molecular Evolution and Adaptive Radiation*. Cambridge: Cambridge University Press.

Glantz, M. H. (2001) *Currents of Change: Impacts of El Niño and La Niña on Climate and Society*, 2nd edn. Cambridge: Cambridge University Press.

Glantz, M. H., Katz, R. W., Nicholl, N (eds) (1991) *Teleconnections Linking Worldwide Climate Anomalies: Scientific Basis and Societal Impact*. Cambridge: Cambridge University Press.

Gleason, H. A. (1926) The individualistic concept of the plant association. *Bulletin of the Torrey Botanical Club* 53, 7–26.

Godbout, J., Fazekas, A., Newton, C. H., Yeh, F. C., and Bousquet, J. (2008) Glacial vicariance in the Pacific Northwest: evidence from a lodgepole pine mitochondrial DNA minisatellite for multiple genetically distinct and widely separated refugia. *Molecular Ecology* 17, 2463–75.

Goudie, A. S. (2005) The drainage of Africa since the Cretaceous. *Geomorphology* 67, 437–56.

Gould, S. J. (1965) Is uniformitarianism necessary? *American Journal of Science* 263, 223–28.

—— (1977) Eternal metaphors in palaeontology. In A. Hallam (ed.) *Patterns in Evolution, as Illustrated by the Fossil Record* (Developments in Palaeontology and Stratigraphy 5), pp. 1–26. Amsterdam: Elsevier.

—— (1984) Toward the vindication of punctuational change. In W. A. Berggrenand and J. A. van Couvering (eds), *Catastrophes and Earth History: the New Uniformitarianism*, pp. 9–34. Princeton, NJ: Princeton University Press.

Graham, R. W. (1979) Paleoclimates and late Pleistocene faunal provinces in North America. In R. L. Humphrey and D. J. Stanford (eds) *Pre-Llano Cultures of the Americas: Paradoxes and Possibilities*, pp. 46–69. Washington, DC: Anthropological Society of Washington.

—— (2005) Quaternary mammal communities: relevance of the individualistic response and non-analogue faunas. *Paleontological Society Papers* 11, 141–58.

Grant, P. R. (1999) *Ecology and Evolution of Darwin's Finches*, 2nd edn. With a new Foreword by Jonathan Weiner. Princeton, NJ: Princeton University Press.

Grant, V. (1977) *Organismic Evolution*. San Francisco, CA: W. H. Freeman.

Gray, M. (2003) *Geodiversity: Valuing and Conserving Abiotic Nature*. Chichester: John Wiley & Sons.

Gregory, K. J. and Walling, D. E. (1976) *Drainage Basin Form and Process: A Geomorphological Approach*, new edition. London: Hodder Arnold.

Gribbin, J. (2004) *Deep Simplicity: Chaos, Complexity and the Emergence of Life*. London: Allen Lane.

Grinnell, J. (1917) The niche-relationships of the California thrasher. *The Auk* 34, 427–33.

Groffman, P., Baron, J., Blett, T., Gold, A., Goodman, I., Gunderson, L., Levinson, B., Palmer, M., Paerl, H., Peterson, G., Poff, N. L., Rejeski, D., Reynolds, J., Turner, M., Weathers, K., and Wiens, J. (2006) Ecological thresholds: the key to successful environmental management or an important concept with no practical application? *Ecosystems* 9, 1–13.

Grove, R. H. (1998) Global impact of the 1789–93 El Niño. *Nature* 393, 318–19.

Grunwald, S. and Reddy K. R. (2008) Spatial behavior of phosphorus and nitrogen in a subtropical wetland. *Soil Science Society of America, Journal* 72, 1174–83.

Gunderson, L. and Holling, C. (eds) (2002) *Panarchy: Understanding Transformations in Human and Natural Systems*. Washington, DC: Island Press.

Gunderson, L. H. and Pritchard, L. (2002) *Resilience and the Behaviour of Large-scale Systems*. Washington, DC: Island Press.

Gutiérrez, R. J. and Harrison, S. (1996) Applications of metapopulation theory to spotted owl management: a history and critique. In D. McCullough (ed.)

Metapopulations and Wildlife Conservation Management, pp. 167–185. Covelo, CA: Island Press.

Hack, J. T. (1960) Interpretation of erosional topography in humid temperate regions. *American Journal of Science* (Bradley Volume) 258-A, 80–97.

Hack, J. T. and Goodlett, J. C. (1960) *Geomorphology and Forest Eecology of a Mountain Region in the Central Appalachians* (US Geological Survey Professional Paper 347). Reston, VA: US Geological Survey.

Hackett, S. J., Kimball, R. T., Reddy, S., Bowie, R. C. K., Braun, E. L., Braun, M. J., Chojnowski, J. L., Cox, W. A., Han, K.-L., Harshman, J., Huddleston, C. J., Marks, B. D., Miglia, K. J., Moore, W. S., Sheldon, F. H., Steadman, D. W., Witt, C. C., and Yuri, T. (2008) A phylogenetic study of birds reveals their evolutionary history. *Science* 320, 1763–68.

Haffer, J. (1969) Speciation in Amazonian forest birds. *Science* 165, 131–37.

Haila, Y. (2002) A conceptual genealogy of fragmentation research: from island biogeography to landscape ecology. *Ecological Applications* 12, 321–34.

Haines-Young, R. H. and Petch, J. R. (1983) Multiple working hypotheses: equifinality and the study of landforms. *Transactions of the Institute of British Geographers*, New Series 8, 458–66.

Hairston, N. G., Jr, Ellner, S. P., Geber, M. A., Yoshida, T., and Fox, J. A. (2005) Rapid evolution and the convergence of ecological and evolutionary time. *Ecology Letters* 8, 1114–27.

Hall, A. M. (1991) Pre-Quaternary landscape evolution in the Scottish Highlands. *Transactions of the Royal Society of Edinburgh: Earth Sciences* 82, 1–26.

Hallam, A. (1973) *A Revolution if the Earth Sciences: From Continental Drift to Plate Tectonics*. Oxford: Clarendon Press.

Hancock, P. L. and Williams, G. D. (1986) Neotectonics. *Journal of the Geological Society, London* 143, 325–26.

Hanski, I. (1986) Population dynamics of shrews on small islands accord with the equilibrium model. *Biological Journal of the Linnean Society* 28, 23–36.

Hanksi, I. (1999) *Metapopulation Ecology*. Oxford: Oxford University Press.

Hanski, I., Pakkala, T., Kuussaari, M., and Guangchun Lei, (1995) Metapopulation persistence of an endangered butterfly in a fragmented landscape. *Oikos* 72, 21–28.

Hare, F. K. (1996) Climatic variation and global change. In I. Douglas, R. J. Huggett, and M. E. Robinson (eds) *Companion Encyclopedia of Geography*, pp. 482–507. London: Routledge.

Harrison, S. (2001) On reductionism and emergence in geomorphology. *Transactions of the Institute of British Geographers*, New Series 26, 327–39.

Harrison, S., Murphy, S. D., and Ehrlich, P. R. (1988) Distribution of the Bay checkerspot butterfly, *Euphydryas editha bayensis*: evidence for a metapopulation model. *American Naturalist* 132, 360–82.

Hays, J. D., Imbrie, J., and Shackelton, N. J. (1976) Variations in the Earth's orbit: pacemaker of the ice ages. *Science* 194, 1121–32.

Head, L. (2007) Evolving nature–culture relationships. In I. Douglas, R. Huggett, and C. Perkins (eds) *Companion Encyclopedia of Geography: From Local to Global*, 2nd edn, pp. 835–46. London and New York: Routledge.

Hergarten, S. and Neugebauer, H. J. (2001) Self-organized critical drainage. *Physical Review Letters* 86, 2689–92.

Hett, J. M. and O'Neill, R. V. (1974) Systems analysis of the Aleut ecosystem. *Arctic Anthropology* 11, 31–40.

Hilty, J. A., Lidicker, W. Z., and Merenlender, A. M. (2006) *Corridor Ecology: The Science and Practice of Linking Landscapes for Biodiversity Conservation*. New York: Island Press.

Hobbs, R. J., Arico, S., Aronson, J., Baron, J. S., Bridgewater, P., Cramer, V. A., Epstein, P. R., Ewel, J. J., Klink, C. A., Lugo, A. E., Norton, D., Ojima, D., Richardson, D. M., Sanderson, E. W., Valladares, F., Vilà, M., Zamora, R., and Zobel, M. (2006) Novel ecosystems: theoretical and management aspects of the new ecological world order. *Global Ecology and Biogeography* 15, 1–7.

Hodkinson, I. D., Coulson, S. J., Webb, N. R. (2004) Invertebrate community assembly along proglacial chronosequences in the high Arctic. *Journal of Animal Ecology* 73, 556–68.

Hole, F. D. (1961) A classification of pedoturbation and some other processes and factors of soil formation in relation to isotropism and anisotropism. *Soil Science* 91, 375–77.

Holling, C. S. (1973) Resilience and stability of ecological systems. *Annual Review of Ecology and Systematics* 4, 1–23.

Horn, H. S. (1981) Succession. In R. M. May (ed.) *Theoretical Ecology: Principles and Applications*, 2nd edn, pp. 253–71. Oxford: Blackwell Scientific Publications.

Houghton, J. T. (2004) *Global Warming: The Complete Briefing*, 3rd edn. Cambridge: Cambridge University Press.

Houghton, J. T., Ding, Y., Griggs, D. J., Noquet, M., van der Linden, J. P., Dai, X., Maskell, K., and Johnson, C. A. (eds) (2001) *Climate Change 2001: The Scientific Basis: Contribution of Working Group I to the Third Assessment Report of the Intergovernmental Panel on Climate Change: The Scientific Basis*. Cambridge: Cambridge University Press and the Intergovernmental Panel on Climate Change.

Huggett, A. (2005) The concept and utility of 'ecological thresholds' in biodiversity conservation. *Biological Conservation* 124, 301–10.

Huggett, R. J. (1973) *Soil Landscape Systems: Theory and Field Evidence*, unpublished PhD Thesis, University of London.

—— (1975) Soil landscape systems: a model of soil genesis. *Geoderma* 13, 1–22.

—— (1985) *Earth Surface Systems* (Springer Series in Physical Environment 1). Heidelberg: Springer-Verlag.

—— (1988) Dissipative system: implications for geomorphology. *Earth Surface Processes and Landforms* 13, 45–49.

—— (1989) *Cataclysms and Earth History: The Development of Diluvialism*. Oxford: Clarendon Press.

—— (1990) *Catastrophism: Systems of Earth History*. London: Edward Arnold.

—— (1991) *Climate, Earth Processes and Earth History*. Heidelberg: Springer.

—— (1995) *Geoecology: An Evolutionary Approach*. London: Routledge.

—— (1997a) *Environmental Change: The Evolving Ecosphere*. London: Routledge.

—— (1997b) *Catastrophism: Asteroid, Comets, and Other Dynamic Events in Earth History*. London: Verso.

—— (2004) *Fundamentals of Biogeography*, 2nd edn. London: Routledge.

—— (2006) *The Natural History of the Earth: Debating Long-term Change in the Geosphere and Biosphere*. Routledge: London.

—— (2007a) *Fundamentals of Geomorphology*, 2nd edn. London: Routledge.

—— (2007b) Drivers of global change. In I. Douglas, R. Huggett, and C. Perkins (eds) *Companion Encyclopedia of Geography: From Local to Global*, pp. 75–91. Abingdon: Routledge.

—— (2007c) Climate. In I. Douglas, R. Huggett, and C. Perkins (eds) *Companion Encyclopedia of Geography: From Local to Global*, pp. 109–28. Abingdon: Routledge.

Huggett, R. J. and Cheesman, J. E. (2002) *Topography and the Environment.* Harlow, Essex: Prentice Hall.

Humphries, C. J. (2000) Form, space and time; which come first? *Journal of Biogeography* 27, 11–15.

Hunt, W. G. and Selander, R. K. (1973) Biochemical genetics of hybridization in European house mice. *Heredity* 31, 11–33.

Hutchinson, G. E. (1957) Concluding remarks. *Cold Spring Harbor Symposia on Quantitative Biology* 22, 415–27.

Hutton, J. (1788) Theory of the Earth; or, an investigation of the laws observable in the composition, dissolution, and restoration of land upon the globe. *Transactions of the Royal Society of Edinburgh* 1, 209–304.

Huxley, J. (1942) *Evolution: The Modern Synthesis.* London: George Allen & Unwin.

Huxley, J. S. (1953) *Evolution in Action.* London: Chatto & Windus.

Hylander, L. D., Silva, E. C., Oliveira, L. J., Silva, S. A., Kuntze, E. K., and Silva, D. X. (1994) Mercury levels in Alto Pantanal: a screening study. *Ambio* 23, 478–84.

Illies, J. (1974) *Introduction to Zoogeography.* Translated by W. D. Williams. London: Macmillan.

Imbrie, J. and Imbrie, K. P. (1986) *Ice Ages: Solving the Mystery.* Cambridge, MA and London: Harvard University Press.

Ingham, D. S. and Samways, M. J. (1996) Application of fragmentation and variegation models to epigaeic invertebrates in South Africa. *Conservation Biology* 10, 1353–58.

Ivanov, B. A. and Melosh, H. J. (2003) Impacts do not initiate volcanic eruptions: eruptions close to the crater. *Geology* 31, 869–72.

Jacobson, M. C., Charlson, R. J., Rodhe, H., and Orians, G. H. (2000) *Earth System Science: From Biogeochemical Cycles to Global Changes.* London and San Diego, CA: Elsevier Academic Press.

Jenny, H. (1941) *Factors of Soil Formation: A System of Quantitative Pedology.* New York: McGraw-Hill.

—— (1961) Derivation of state factor equations of soils and ecosystems. *Soil Science Society of America Proceedings* 25, 385–88.

—— (1980) *The Soil Resource: Origin and Behaviour* (Ecological Studies, vol. 37). New York: Springer.

Johnson, D. L. (1990) Biomantle evolution and the redistribution of earth materials and artefacts. *Soil Science* 149, 84–102.

—— (1993a) Dynamic denudation evolution of tropical, subtropical and temperate landscapes with three tiered soils: toward a general theory of landscape evolution. *Quaternary International* 17, 67–78.

—— (1993b) Biomechanical processes and the Gaia paradigm in a unified pedo-geomorphic and pedo-archaeologic framework: dynamic denudation. In J. E. Foss, M. E. Timpson, and M. W. Morris (eds) *Proceedings of the First International Conference on Pedo-Archaeology* (University of Tennessee Agricultural Experimental Station, Special Paper 93–03), pp. 41–67. Knoxville, TN: University of Tennessee Agricultural Experimental Station.

—— (1994) Reassessment of early and modern soil horizon designation frameworks as associated pedogenetic processes: are midlatitude A E B–C horizons equivalent to tropical M S W horizons?' *Soil Science (Trends in Agricultural Science)* 2, 77–91.

—— (2002) Darwin would be proud: bioturbation, dynamic denudation, and the power of theory in science. *Geoarchaeology: An International Journal* 17, 7–40.

Johnson, D. L. and Hole, F. D. (1994) Soil formation theory: a summary of its principal impacts on geography, geomorphology, soil–geomorphology, Quaternary geology and paleopedology. In R. Amundson (ed.) *Factors of Soil Formation: A Fiftieth Anniversary Retrospective* (Soil Science Society of America Special Publication 33), pp. 111–26. Madison, WI: Soil Science Society of America.

Johnson, D. L., Domier, J. E. J., and Johnson, D. N. (2005) Animating the biodynamics of soil thickness using process analysis: a dynamic denudation approach to soil formation. *Geomorphology* 67, 23–46.

Johnson, E. A. and Miyanishi, K. (eds) (2007) *Plant Disturbance Ecology: the Process and the Response*. Burlington MA: Elsevier Academic Press.

Johnson, R. L. (2006) *Plate Tectonics*. Minneapolis, MN: Twenty-First Century Books.

Jones, D. K. C. (1999) Evolving models of the Tertiary evolutionary geomorphology of southern England, with special reference to the Chalklands. In B. J. Smith, W. B. Whalley, and P. A. Warke (eds) *Uplift, Erosion and Stability: Perspectives on Long-term Landscape Development* (Geological Society, London, Special Publication 162), pp. 1–23. London: The Geological Society.

Karanth, K. U. and Stith, B. M. (1999) Prey depletion as a critical determinant of tiger population viability. In J. Seidensticker, S. Christie, and P. Jackson (eds) *Riding the Tiger: Tiger Conservation in Human-dominated Landscapes*, pp. 100–113. London: The Zoological Society of London; Cambridge: Cambridge University Press.

Karlstrom, E. T. and Osborn, G. (1992) Genesis of buried paleosols and soils in Holocene and late Pleistocene tills, Bugaboo Glacier area, British Columbia, Canada. *Arctic and Alpine Research* 24, 108–23.

Kettlewell, H. B. D. (1973) *The Evolution of Melanism: The Study of a Recurring Necessity, with Special Reference to Industrial Melanism in Lepidoptera*. Oxford: Clarendon Press.

Kirchner, J. W. (1991) The Gaia hypotheses: are they testable? Are they useful? In S. H. Schneider and P. J. Boston (eds) *Scientists on Gaia*, pp. 38–46. Cambridge, MA and London: MIT Press.

Kitayama, K., Mueller-Dombois, D., and Vitousek, P. M. (1995) Primary succession of Hawaiian montane rain forest on a chronosequence of eight lava flows. *Journal of Vegetation Science* 6, 211–22.

Kleidon, A. (2002) Testing the effect of life on Earth's functioning: how Gaian is the Earth System? *Climatic Change* 52, 383–89.

—— (2004) Beyond Gaia: thermodynamics of life and Earth system functioning. *Climatic Change* 66, 271–319.

—— (2007) Thermodynamics and environmental constraints make the biosphere predictable – a response to Volk. *Climatic Change* 85, 259–66.

Koepfli, K.-P., Deere, K. A., Slater, G. J., Begg, C., Begg, K., Grassman, L., Lucherini, M., Veron, G., and Wayne, R. K. (2008) Multigene phylogeny of the Mustelidae: resolving relationships, tempo and biogeographic history of a mammalian adaptive radiation. *BMC Biology* 6, 10.

Kukla, G. and Gavin, J. (2004) Milankovitch climate reinforcements. *Global and Planetary Change* 40, 27–48.

Kumazawa, M. and Maruyama, S. (1994) Whole earth tectonics. *Journal of the Geological Society of Japan* 100, 81–102.

Kump. L. R., Kasting, J. F., and Crane, R. G. (2004) *The Earth System: An Introduction to Earth System Science*, 2nd edn. Upper Saddle River, NJ: Pearson Education.

Ladle, R. J. and Malhado, A. C. M. (2007) Responding to biodiversity loss. In I. Douglas, R. Huggett, and C. Perkins (eds) *Companion Encyclopedia of Geography: From Local to Global*, 2nd edn, pp. 821–34. London and New York: Routledge.

Laity, J. (2008) *Deserts and Desert Environments*. Chichester: John Wiley & Sons.

Larson, R. L. (1991) Latest pulse of the Earth: evidence for a mid-Cretaceous superplume. *Geology* 19, 547–50.

Lawton, J. H. and May, R. M. (eds) (1995) *Extinction Rates*. Oxford: Oxford University Press.

Laycock, A. H. (1987) The amount of Canadian water and its distribution. In M. C. Healey and R. R. Wallace (eds) *Canadian Aquatic Resources* (Canadian Bulletin of Fisheries and Aquatic Sciences 215), pp. 13–42. Ottawa: Department of Fisheries and Oceans.

Legrand, J. P., Le Goff, M., Mazaudier, C., and Schröder, W. (1992) Solar and auroral activities during the seventeenth century. In W. Schröder and J. P. Legrand (eds) *Solar–Terrestrial Variability and Global Change* (Selected Papers from the Symposia of the Interdivisional Commission on History of the IAGA during the IUGG/IAGA Assembly, held in Vienna, 1991), pp. 40–76. Bremen–Roennebeck, Germany: Interdivisional Commission on History of the International Association of Geomagnetism and Aeronomy (IAGA).

Lenton, T., Held, H., Kriegler, E., Hall, J., Lucht, W., Rahmstorf, S., and Schellnhuber, H. J. (2008) Tipping elements in the Earth's climate system. *Proceedings of the National Academy of Sciences* 105, 1786–93.

Lévêque, C. and Mounolou, J.-C. (2003) *Biodiversity*. Chichester: John Wiley & Sons.

Levins, R. (1969) Some demographic and genetic consequences of environmental heterogeneity for biological control. *Bulletin of the Entomological Society of America* 15, 237–40.

—— (1970) Extinction. In M. Gerstenhaber (ed.) *Some Mathematical Questions in Biology*, pp. 77–107. Providence, RI: American Mathematical Society.

Lidmar-Bergström, K., Ollier, C. D., and Sulebak, J. R. (2000) Landforms and uplift history of southern Norway. *Global and Planetary Change* 24, 211–31.

Liebig, J. (1840) *Organic Chemistry and its Application to Agriculture and Physiology*, English edn edited by L. Playfair and W. Gregory. London: Taylor & Walton.

Lindeman, R. L. (1942) The trophic–dynamic aspect of ecology. *Ecology* 23, 399–418.

Lindenmayer, D. B. and Fischer, J. (2006a) Tackling the habitat fragmentation panchreston. *Trends in Ecology and Evolution* 22, 127–32.

—— (2006b) *Habitat Fragmentation and Landscape Change: An Ecological and Conservation Synthesis*. Washington, DC: Island Press.

Linton, D. L. (1955) The problem of tors. *Geographical Journal* 121, 289–91.

Litaor, M. I., Barth, G., Zika, E. M., Litus, G., Moffitt, J. and Daniels, H. (1998)

The behavior of radionuclides in the soils of Rocky Flats, Colorado. *Journal of Environmental Radioactivity* 38, 17–46.

Lomborg, B. (2007) *Cool It: The Skeptical Environmentalist's Guide to Global Warming.* London: Cyan and Marshall Cavendish.

Lomolino, M. V. (1986) Mammalian community structure on islands: the importance of immigration, extinction and integrative effects. *Biological Journal of the Linnean Society* 28, 1–21.

—— (2000a) Ecology's most general, yet protean pattern: the species–area relationship. *Journal of Biogeography* 27, 17–26.

—— (2000b) A species-based theory of insular biogeography. *Global Ecology & Biogeography* 9, 39–58.

Lomolino, M. V. and Weiser, M. D. (2001) Towards a more general species–area relationship: diversity on all islands, great and small. *Journal of Biogeography* 28, 431–45.

Lorenz, E. N. (1963a) Deterministic nonperidic flow. *Journal of Atmospheric Sciences* 20, 130–41.

—— (1963b) Atmosphere models as dynamic systems. In M. F. Shlesinger, R. Cawley, A. W. Saenz, and W. Zachary (eds) *Perspectives in Nonlinear Dynamics*, pp. 1–17. Singapore: World Scientific Publishing Company.

Losos, J. B. and Glor, R. E. (2003) Phylogenetic comparative methods and the geography of speciation. *Trends in Ecology and Evolution* 18, 220–27.

Lotka A. J. (1924) *Elements of Physical Biology.* Baltimore, MD: Williams & Wilkins.

Loucks, O. (1962) A forest classification for the Maritime Provinces. *Proceedings of the Nova Scotian Institute of Science* 259(2), 85–167, with separate map at 1 inch equals 19 miles.

Lovejoy, T. E. and Hannah, L. (2006) *Climate Change and Biodiversity.* New Haven, CT: Yale University Press.

Lovelock, J. E. (1965) A physical basis for life detection experiments. *Nature* 207, 568–70.

—— (1972) Gaia as seen through the atmosphere. *Atmospheric Environment* 6, 579–80.

—— (1979) *Gaia: A New Look at Life on Earth.* Oxford and New York: Oxford University Press.

—— (1988) *The Ages of Gaia: A Biography of Our Living Earth.* Oxford: Oxford University Press.

—— (1991) Geophysiology – the science of Gaia. In S. H. Schneider and P. J. Boston (eds) *Scientists on Gaia*, pp. 3–10. Cambridge, MA: MIT Press.

—— (2000) *Homage to Gaia: the Life of an Independent Scientist.* Oxford: Oxford University Press.

—— (2003) The living Earth. *Nature* 426, 769–70.

Lovelock, J. E. and Margulis, L. (1974) Atmospheric homeostasis by and for the biosphere: the Gaia hypothesis. *Tellus* 26, 2–10.

Lundelius, E. L., Jr, Graham, R. W., Anderson, E., Guilday, J., Holman, J. A., Steadman, D., and Webb, S. D. (1983) Terrestrial vertebrate faunas. In S. C. Porter (ed.) *Late-Quaternary Environments of the United States. Vol. 1. The Late Pleistocene*, pp. 311–53. London: Longman.

Luo, Y., Wan, S., Hui, D., and Wallace, L. L. (2001) Acclimatization of soil respiration to warming in a tall grass prairie. *Nature* 413, 622–25.

Lyell, C. (1830–33) *Principles of Geology, Being an Attempt to Explain the Former*

Changes of the Earth's Surface, by Reference to Causes Now in Operation. 3 vols. London: John Murray.

—— (1830–33) *Principles of Geology. First Edition.* A facsimile edition, with a new introduction by Martin S. Rudwick. 3 vols. Chicago, IL and London: The University of Chicago Press.

Lyons, K, Smith, F. A., Wagner, P. J., White, E. P., and Brown, J. H. (2004). Was a 'hyperdisease' responsible for the late Pleistocene megafaunal extinction? *Ecology Letters* 7, 859–68.

MacArthur, R. H. and Wilson, E. O. (1963) An equilibrium theory of insular zoogeography. *Evolution* 17, 373–87.

—— (1967) *The Theory of Island Biogeography.* Princeton, NJ: Princeton University Press.

MacPhee, R. D. and Marx, P. A. (1997) The 40,000 year plague: humans, hyperdisease, and first-contact extinctions. In S. A. Goodman and B. D. Patterson (eds) *Natural Change and Human Impact in Madagascar*, pp. 169–217. Washington, DC: Smithsonian Institution Press.

Major, J. (1951) A functional factorial approach to plant ecology. *Ecology* 32, 392–412.

Mannion, A. M. (1997) *Global Environmental Change: A Natural and Cultural Environmental History*, 2nd edn. Harlow, Essex: Longman.

—— (1999) *Natural Environmental Change.* London: Routledge.

Marshall, P. and Schuttenberg, H. (2006) *A Reef Manager's Guide to Coral Bleaching.* Townsville, Australia: Great Barrier Reef Authority Marine Park.

Maruyama, S. (1994) Plume tectonics. *Journal of the Geological Society of Japan* 100, 24–49.

Maruyama, S., Kumazawa, M., and Kawakami, S. (1994) Towards a new paradigm on the Earth's dynamics. *Journal of the Geological Society of Japan* 100, 1–3.

Mason, H. L. (1954) Migration and evolution in plants. *Madroño* 12, 161–92.

Mayr, E. (1942) *Systematics and the Origin of Species.* New York: Columbia University Press.

—— (1970) *Population, Species, and Evolution* (An abridgement of *Animal Species and Evolution*). Cambridge, MA and London: The Belknap Press of Harvard University Press.

McGlone, M. S. (2005) Goodbye Gondwana. *Journal of Biogeography* 32, 739–40.

McIntyre, S. and Barrett, G. W. (1992) Habitat variegation, an alternative to fragmentation. *Conservation Biology* 6, 146–47.

McManus, J. W. and Polsenberg, J. F. (2004) Coral–algal phase shifts on coral reefs: ecological and environmental aspects. *Progress in Oceanography* 60, 263–79.

McSweeney, K., Slater, B. K., Hammer, R. D., Bell, J. C., Gessler, P. E., and Petersen, G. W. (1994) Towards a new framework for modeling the soil–landscape continuum. In R. Amundson, J. Harden, and M. Singer (eds) *Factors of Soil Formation: A Fiftieth Anniversary Retrospective* (Soil Science Society of America Special Publication Number 33), pp. 127–45. Madison, WI: Soil Science Society of America.

Meigs, P. (1953) World distribution of arid and semi-arid homoclimates. In *Review of Research on Arid Zone Hydrology, Arid Zone Programme 1*, pp. 203–10. Paris: UNESCO.

Middleton, N. J. and Thomas, D. S. G. (eds) (1997) *World Atlas of Desertification*, 2nd edn. London: Arnold.

—— (1997) *World Atlas of Desertification*, 2nd edn. London: Arnold.

Midgley, G. F., Hannah, L., Millar, D., Rutherford, M. C., and Powrie, L. W. (2002) Assessing the vulnerability of species richness to anthropogenic climate change in a biodiversity hotspot. *Global Ecology & Biogeography* 11, 445–51.

Millennium Ecosystem Assessment (2005) *Ecosystems and Human Well-being: Desertification Synthesis*. Washington, DC: World Resources Institute.

Milne, G. (1935a) Some suggested units of classification and mapping, particularly for East African soils. *Soil Research* 4, 183–98.

—— (1935b) Composite units for the mapping of complex soil associations. *Transactions of the Third International Congress of Soil Science, Oxford, England, 1935* 1, 345–47.

Milner, R. (1990) *The Encyclopedia of Evolution: Humanity's Search for Its Origins*. Foreword by Stephen Jay Gould. New York and Oxford: Facts on File.

Milton, S. J. (2003) 'Emerging ecosystems': a washing-stone for ecologists, economists, and sociologists? *South African Journal of Science* 99, 404–06.

Mock, K. E., Bentz, B. J., O'Neill, E. M., Chong, J. P., Orwin, J., and Pfrender, M. E. (2007) Landscape-scale genetic variation in a forest outbreak species, the mountain pine beetle (*Dendroctonus ponderosae*). *Molecular Ecology* 16, 553–68.

Mooney, H. A., Mack, R. N., McNeely, J. A., Neville, L. E., Schei. P. J., and Waage, J. (eds) (2005) *Invasive Alien Species: A New Synthesis*. Washington, DC: Island Press.

Moore, I. G., Grayson, R. B., and Ladson, A. R. (1991) Digital terrain modelling: a review of hydrological, geomorphological, and biological applications. *Hydrological Processes* 5, 3–30.

Morgan, W. J. (1971) Convection plumes in the lower mantle. *Nature* 230, 42–43.

Morison, C. G. T. (1949) The catena concept and the classification of tropical soils. In *Proceedings of the First Commonwealth Conference on Tropical and Sub-Tropical Soils, 1948* (Commonwealth Bureau of Soil Science, Technical Communication No. 46). Harpenden, England: Commonwealth Bureau of Soil Science.

Mörner, N.-A. (1980) The northwest European 'sea-level laboratory' and regional Holocene eustasy. *Palaeogeography, Palaeoclimatology, Palaeoecology* 29, 281–300.

—— (1987) Models of global sea-level changes. In M. J. Tooley and I. Shennan (eds) *Sea-Level Changes*, pp. 332–55. Oxford: Basil Blackwell.

—— (1994) Internal response to orbital forcing and external cyclic sedimentary sequences. In P. L. De Boer and D. G. Smith (eds) *Orbital Forcing and Cyclic Sequences* (Special Publication Number 19 of the International Association of Sedimentologists), pp. 25–33. Oxford: Blackwell Scientific Publications.

Morse, S. A. (2000) A double magmatic heat pump at the core–mantle boundary. *American Mineralogist* 85, 1589–94.

Muhs, D. R. (1982) The influence of topography on the spatial variability of soils in Mediterranean climates. In C. E. Thorn (ed.) *Space and Time In Geomorphology*, pp. 269–84. London: George Allen & Unwin.

—— (1984) Intrinsic thresholds in soil systems. *Physical Geography* 5, 99–110.

Naiman, R. J. and Décamps, H. (1997) The ecology of interfaces: riparian zones. *Annual Review of Ecology and Systematics* 28, 621–58.

Namias, J. (1950) The index cycle and its role in the general circulation. *Journal of Meteorology* 17, 130–39.

Napier, W. M. and Clube, S. V. M. (1979) A theory of terrestrial catastrophism. *Nature* 282, 455–59.

Naqvi, S. M., Howell, R. D., and Sholas, M. (1993) Cadmium and lead residues in field-collected red swamp crayfish (*Procambarus clarkii*) and uptake by alligator weed, *Alternanthera philoxiroides*. *Journal of Environmental Science and Health* B28, 473–85.

Naydenov, K., Senneville, S., Beaulieu, J., Tremblay, F., and Bousquet, J. (2007) Glacial vicariance in Eurasia: mitochondrial DNA evidence from Scots pine for a complex heritage involving genetically distinct refugia at mid-northern latitudes and in Asia Minor. *BMC Evolutionary Biology* 7, 233.

Neal, D. (2004) *Introduction to Population Biology*. Cambridge: Cambridge University Press.

Nelson, G. and Rosen, D. E. (eds) (1981) *Vicariance Biogeography: A Critique* (Symposium of the Systematics Discussion Group of the America Museum of Natural History May 2–4, 1979). New York: Columbia University Press.

Niemiller, M. L., Fitzpatrick, B. M., and Miller, B. T. (2008) Recent divergence with gene-flow in Tennessee cave salamanders (Plethodontidae: *Gyrinophilus*) inferred from gene genealogies. *Molecular Ecology* 17, 2258–75.

Nikiforoff, C. C. (1959) Reappraisal of the soil. *Science* 129, 186–96.

Noon, B. R. and Franklin, A. B. (2002) Scientific research and the spotted owl (*Strix occidentalis*): opportunities for major contributions to avian population ecology. *The Auk* 119, 311–20.

Nores, M. (1999) An alternative hypothesis for the origin of Amazonian bird diversity. *Journal of Biogeography* 26, 475–85.

Nosil, P. (2008) Speciation with gene flow could be common. *Molecular Ecology* 17, 2103–6.

Odum, H. T. (1994) *Ecological and General Systems: An Introduction to Systems Ecology*. Niwot, CO: University Press of Colorado.

Ohlemüller, R., Gritti, E. S., Sykes, M. T., and Thomas, C. D. (2006) Towards European climate risk surfaces: the extent and distribution of analogous and non-analogous climates 1931–2000. *Global Ecology and Biogeography* 15, 395–405.

Oke, T. R. (1982) The energetic basis of the urban heat island. *Quarterly Journal of the Royal Meteorological Society* 108, 1–24.

—— (1987) *Boundary Layer Climates*, 2nd edn. London: Routledge.

Oldfield, F. (2005) *Environmental Change: Key Issues and Alternative Approaches*. Cambridge: Cambridge University Press.

Ollier, C. D. (1959) A two-cycle theory of tropical pedology. *Journal of Soil Science* 10: 137–48.

—— (1960) The inselbergs of Uganda. *Zeitschrift für Geomorphologie* NF 4, 470–87.

—— (1967) Landform description without stage names. *Australian Geographical Studies* 5, 73–80.

—— (1968) Open systems and dynamic equilibrium in geomorphology. *Australian Geographical Studies* 6, 167–70.

—— (1981) *Tectonics and Landforms* (Geomorphology Texts 6). London and New York: Longman.

—— (1991) *Ancient Landforms*. London and New York: Belhaven Press.

—— (1992) Global change and long-term geomorphology. *Terra Nova* 4, 312–19.

—— (1995) Tectonics and landscape evolution in southeast Australia. *Geomorphology* 12, 37–44.

—— (1996) Planet Earth. In I. Douglas, R. J. Huggett, and M. E. Robinson (eds) *Companion Encylopedia of Geography*, pp. 15–43. London: Routledge.

—— (2004) The evolution of mountains on passive continental margins. In P. N. Owens and O. Slaymaker (eds) *Mountain Geomorphology*, pp. 59–88. London: Arnold.

—— (2005) A plate tectonic failure: the geological cycle and conservation of continents and oceans. *Annals of Geophysics (Annali di Geofisica)* 48 (Supplement), 961–70.

Ollier, C. D. and Pain, C. F. (1994) Landscape evolution and tectonics in southeastern Australia. *AGSO Journal of Australian Geology and Geophysics* 15, 335–45.

—— (1996) *Regolith, Soils and Landforms*. Chichester: John Wiley & Sons.

—— (1997) Equating the basal unconformity with the palaeoplain: a model for passive margins. *Geomorphology* 19, 1–15.

Oreskes, N. (ed.) (2003) *Plate Tectonics: An Insider's History of the Modern Theory of the Earth*. Boulder, CO: Westview Press.

Paine, A. D. M. (1985) 'Ergodic' reasoning in geomorphology: time for a review of the term? *Progress in Physical Geography* 9, 1–15.

Palumbi, S. R. (2001) *The Evolution Explosion: How Humans Cause Rapid Evolutionary Change*. New York: W.W. Norton.

Paton, T. R., Humphreys, G. S., and Mitchell, P. B. (1995) *Soils: A New Global View*. London: UCL Press.

Pavlides, S. B. (1989) Looking for a definition of neotectonics. *Terra Nova* 1, 233–35.

Penvenne, L. J. (1995) Turning up the heat. *New Scientist* 148 (no. 2008), 26–30.

Phillips, J. D. (1999a) Divergence, convergence, and self-organization in landscapes. *Annals of the Association of American Geographers* 89, 466–88.

—— (1999b) *Earth Surface Systems: Complexity, Order, and Scale*. Oxford: Blackwell.

—— (2001) The relative importance of intrinsic and extrinsic factors in pedodiversity. *Annals of the Association of American Geographers* 91, 609–21.

—— (2006a) Deterministic chaos and historical geomorphology: a review and look forward. *Geomorphology* 76, 109–21.

—— (2006b) Evolutionary geomorphology: thresholds and nonlinearity in landform response to environmental change. *Hydrology and Earth System Sciences* 10, 731–42.

—— (2007) The perfect landscape. *Geomorphology* 84, 159–69.

—— (2008) Goal functions in ecosystem and biosphere evolution. *Progress in Physical Geography* 32, 51–64.

Pitman, A. J. (2005) On the role of Geography in Earth System Science. *Geoforum* 36, 137–48.

Playfair, J. (1802) *Illustrations of the Huttonian Theory of the Earth*. London: Cadell & Davies; Edinburgh: William Creech.

—— (1964) *Illustrations of the Huttonian Theory of the Earth*. A facsimile edition, with an introduction by George W. White. New York: Dover Books.

Poincaré, H. (1881–86) Mémoire sur les courbes définies par une équation différentielle. *Journal des Mathématiques Pures et Appliquées* 3e série 7 (1881), 375–422; 3e série 8 (1882), 251–296; 4e série 1 (1885), 167–244; 4e série 4 (1886), 2, 151–217.

Preston, F. W. (1962) The canonical distribution of commonness and rarity. *Ecology* 43, 185–215, 410–32.

Price, N. J. (2001) *Major Impacts and Plate Tectonics: A Model for the Phanerozoic Evolution of the Earth's Lithosphere.* London: Routledge.

Prigogine, I. (1980) *From Being to Becoming: Time and Complexity in the Physical Sciences.* San Francisco, CA: W. H. Freeman.

Prokoph, A., Rampino, M. R., and El Bilali, H. (2004) Periodic components in the diversity of calcareous plankton and geological events over the past 230 Myr. *Palaeogeography, Palaeoclimatology, Palaeoecology* 207, 105–25.

Purdue, J. R. (1989) Changes during the Holocene in the size of the white-tailed deer (*Odocoileus virginianus*) from central Illinois. *Quaternary Research* 32, 307–16.

Puttker, T. (2008) *Effects of Habitat Fragmentation on Small Mammals of the Atlantic Forest, Brazil.* Saarbrücken: VDM Verlag Dr. Muller Aktiengesellschaft & Co. KG.

Queiroz, A. de (2005) The resurrection of oceanic dispersal in historical biogeography. *Trends in Ecology and Evolution* 20, 68–73.

Rahmstorf, S. (2003) The current climate. *Nature* 421, 699.

Rampino, M. R. (1989) Dinosaurs, comets and volcanoes. *New Scientist* 121, 54–58.

—— (2002) Role of the Galaxy in periodic impacts and mass extinctions on the Earth. In C. Koeberl and K. G. MacLeod (eds) *Catastrophic Events and Mass Extinctions: Impacts and Beyond* (Geological Society of America Special Paper 356), 667–78. Boulder, CO: The Geological Society of America.

Rathgeber, C. B. K., Misson, L., Nicault, A., and Guiot, J. (2005) Bioclimatic model of tree radial growth: application to the French Mediterranean Aleppo pine forests. *Tree – Structure and Function* 19, 162–76.

Raunkiaer, C. (1934) *The Life Forms of Plants and Statistical Plant Geography, Being the Collected Papers of C. Raunkiaer.* Translated by H. Gilbert-Carter and A. G. Tansley. Clarendon Press: Oxford.

Reading, H. G. (ed.) (1978) *Sedimentary Environments and Facies.* Oxford: Blackwell.

Rees, W. E. (1995) Achieving sustainability: reform or transformation? *Journal of Planning Literature* 9, 343–61.

Reid, W. V. and Miller, K. R. (1989) *Keeping Options Alive: The Scientific Basis for Conserving Biodiversity.* Washington, DC: World Resources Institute.

Renwick, W. H. (1992) Equilibrium, disequilibrium, and non-equilibrium landforms in the landscape. *Geomorphology* 5, 265–76.

Retallack, G. J. (1986) The fossil record of soils. In V. P. Wright (ed.) *Palaeosols: Their Recognition and Interpretation*, pp. 1–57 Oxford: Blackwell Scientific.

—— (1990) *Soils of the Past: An Introduction to Paleopedology*, 1st edn. Boston: Unwin Hyman.

—— (2001) *Soils of the Past: An Introduction to Paleopedology*, 2nd edn. Oxford: Blackwell.

—— (2003) Soils and global change in the carbon cycle over geological time. In J. I. Drever (ed.) and H. D. Holland and K. K. Turekian (executive eds) *Treatise on Geochemistry*, Vol. 5, pp. 581–605. Amsterdam: Elsevier.

Rhoads, B. L. (2006) The dynamic basis of geomorphology reenvisioned. *Annals of the Association of American Geographers* 96, 14–30.

Rhodes II, R. S. (1984) Paleoecological and regional paleoclimatic implications of the Farmdalian Craigmile and Woodfordian Waubonsie mammalian local faunas, southwestern Iowa. *Illinois State Museum Report of Investigations* 40, 1–51.

Richards, A. E. (2002) Complexity in physical geography. *Geography* 87, 99–107.

Richardson, D. M. and van Wilgen, B. W. (1992) Ecosystem, community and species response to fire in mountain fynbos: conclusions from the Swartbosk-loof experiment. In B. W. van Wilgen, D. M. Richardson, F. J. Kruger, and H. J. van Hensbergen (eds) *Fire in South African Mountain Fynbos: Ecosystem, Community and Species Response at Swartboskloof* (Ecological Studies, vol. 93), pp. 273–84. New York: Springer.

Ridley, M. (2003) *Evolution*, 3rd edn. Oxford: Blackwell Science.

Riehl, H. (1954) *Tropical Meteorology*. New York and London: McGraw-Hill.

Rivas, V., Cendrero, A., Hurtado, M., Cabral, M., Giménez, J., Forte, L., del Río, L., Cantú, M., and Becker, A. (2006) Geomorphic consequences of urban development and mining activities; an analysis of study areas in Spain and Argentina. *Geomorphology* 73, 185–206.

Rohde, R. A. and Muller, R. A. (2005) Cycles in fossil diversity. *Nature* 434, 208–10.

Root, R. B. (1967) The niche exploitation pattern of the blue-gray gnatcatcher. *Ecological Monographs* 37, 317–50.

Rose, M. R. and Lauder, G. V. (eds) (1996) *Adaptation*. San Diego: Academic Press.

Roughgarden, J., May, R. M., and Levin, S. A. (eds) (1989) *Perspectives in Ecological Theory*. Princeton, NJ: Princeton University Press.

Roy, A. G., Jarvis, R. S., and Arnett, R. R. (1980) Soil-slope relationships within a drainage basin. *Annals of the Association of American Geographers* 70, 397–412.

Rudwick, M. J. S. (1992) Darwin and catastrophism. In J. Bourriau (ed.) *Understanding Catastrophe*, pp. 57–58. Cambridge: Cambridge University Press.

Rustad, L. E. (2001) Matter of time on the prairie. *Nature* 413, 578–79.

Rykiel, E. J., Jr, Coulson, R. N., Sharpe, P. J. H., Allen, T. F. H., and Flamm, R. O. (1988) Disturbance propagation by bark beetles as an episodic landscape phenomenon. *Landscape Ecology* 1, 129–39.

Ryrholm, N. (1988) An extralimital population in a warm climatic outpost: the case of the moth *Idaea dilutaria* in Scandinavia. *International Journal of Biometeorology* 32, 205–16.

Savigear, R. A. G. (1952) Some observations on slope development in South Wales. *Transactions of the Institute of British Geographers* 18, 31–52.

—— (1956) Technique and terminology in the investigations of slope forms. In *Premier Rapport de la Commission pour l'Étude des Versants*, pp. 66–75. Amsterdam: Union Géographique Internationale.

Savolainen, V., Anstett, M. C., Lexer, C., Hutton, I., Clarkson, J. J., Norup, M. V., Powell, M. P., Springate, D., Salamin, N., and Baker, W. J. (2006) Sympatric speciation in palms on an oceanic island. *Nature* 441: 210–13.

Sayre, N. F. (2008) The genesis, history, and limits of carrying capacity. *Annals of the Association of American Geographers* 98, 120–34.

Schaetzl, R. J. and Anderson, S. (2005) *Soils: Genesis and Geomorphology*. Cambridge: Cambridge University Press.

Scheidegger, A. E. (1979) The principle of antagonism in the Earth's evolution. *Tectonophysics* 55, T7–T10.

—— (1983) Instability principle in geomorphic equilibrium. *Zeitschrift für Geomorphologie* NF 27, 1–19.

—— (1986) The catena principle in geomorphology. *Zeitschrift für Geomorphologie*, NF 30, 257–73.

Schluter, D. (2000) *The Ecology of Adaptive Radiation*. Oxford: Oxford University Press.

Schumm, S. A. (1956) Evolution of drainage systems and slopes in badlands at Perth Amboy, New Jersey. *Bulletin of the Geological Society of America* 67, 597–646.

—— (1963) Sinuosity of alluvial rivers on the Great Plains. *Bulletin of the Geological Society of America* 74, 1089–100.

—— (1979) Geomorphic thresholds: the concept and its applications. *Transactions of the Institute of British Geographers* New Series 4, 485–515.

Schumm, S. A. and Lichty, R. W. (1965) Time, space and causality in geomorphology. *American Journal of Science* 263, 110–19.

Schwilk, D. and Ackerly, D. (2001) Flammability and serotiny as strategies: correlated evolution in pines. *Oikos* 94, 326–36.

Shafer, C. L. (1990) *Nature Reserves: Island Theory and Conservation Practice*. Washington, DC and London: Smithsonian Institution Press.

Shaw, H. R. (1994) *Craters, Cosmos, Chronicles: A New Theory of the Earth*. Stanford, CA: Stanford University Press.

Shelford, V. E. (1911) Physiological animal geography. *Journal of Morphology* 22, 551–618.

Shiklomanov, I. A. and Rodda, J. C. (eds) (2003) *World Water Resources at the Beginning of the Twenty-First Century (International Hydrology) (International Hydrology Series)*. Cambridge: Cambridge University Press.

Simpson, G. G. (1944) *Tempo and Mode in Evolution*. New York: Columbia University Press.

Sinsch, U. (1992) Structure and dynamic of a natterjack toad metapopulation (*Bufo calamita*). *Oecologia* 90, 489–99.

Sjögren, P. (1991) Extinction and isolation gradients in metapopulations: the case of the pool frog (*Rana lessonae*). *Biological Journal of the Linnean Society* 42, 135–47.

Skórka, P., Martyka, R., and Wójcik, J. D. (2006) Species richness of breeding birds at a landscape scale: which habitat type is the most important? *Acta Ornithologica* 41, 49–54.

Slobodkin, L. B. (1961) *The Growth and Regulation of Animal Populations*. New York, London: Holt, Rinehart and Winston.

Smith, A. G., Smith, D. G., and Funnell, B. M. (1994) *Atlas of Mesozoic and Cenozoic Coastlines*. Cambridge: Cambridge University Press.

Soil Survey Staff (1975) *Soil Taxonomy: A Basic System of Soil Classification for Making and Interpreting Soil Surveys* (US Department of Agriculture, Agricultural Handbook 436). Washington, DC: US Government Printing Office.

—— (1999) *Soil Taxonomy: A Basic System of Soil Classification for Making and Interpreting Soil Surveys*, 2nd edn (US Department of Agriculture, Natural Resources Conservation Service, Agricultural Handbook 436). Washington, DC: US Government Printing Office.

Soon, W. W.-H. and Yaskell, S. H. (2003) *The Maunder Minimum and the Variable Sun–Earth Connection*. River Edge, NJ: World Scientific Publishing.

Stablein, G. (1984) Geomorphic altitudinal zonation in the Arctic–alpine mountains of Greenland. *Mountain Research and Development* 4, 319–31.

Steel, D. I. (1991) Our asteroid-pelted planet. *Nature* 354, 265–67.

—— (1995) *Rogue Asteroids and Doomsday Comets: The Search for the Million Megaton Menace That Threatens Life on Earth*. Foreword by Arthur C. Clarke. New York: John Wiley & Sons.

Steel, D. I., Asher, D. J., Napier, W. M., and Clube, S. V. M. (1994) Are impacts correlated in time? In T. Gehrels (ed.), with the editorial assistance of M. S. Matthews and A. M. Schumann, *Hazards Due to Comets and Asteroids*, pp. 463–77. Tucson, AZ and London: The University of Arizona Press.

Steffen, W., Sanderson, A., Tyson, P. D., Jäger, J., Matson, P. A., Moore III, B., Oldfield, F., Richardson, K., Schellnhuber, H. J., Turner II, B. L., and Wasson, R. J. (eds) (2004) *Global Change and the Earth System: A Planet Under Pressure* (Global Change: the IGBP Series). Berlin, Heidelberg: Springer-Verlag.

Stewart, I. (1997) *Does God Play Dice? The New Mathematics of Chaos*, new edn. Harmondsworth: Penguin Books.

Strahler, A. N. (1952) Dynamic basis of geomorphology. *Bulletin of the Geological Society of America* 63, 923–38.

—— (1980) Systems theory in physical geography. *Physical Geography* 1, 1–27.

Summerfield, M. A. (1991) *Global Geomorphology: An Introduction to the Study of Landforms*. Harlow, Essex: Longman.

Sutherland, J. P. (1974) Multiple stable states in natural communities. *American Naturalist* 108, 859–73.

Sutherland, R. A., van Kessel, C., Farrell, R. E., and Pennock, D. J. (1993) Landscape-scale variations in soil nitrogen-15 natural abundance. *Soil Science Society of America Journal* 57, 169–78.

Swanson, D. K. (1985) Soil catenas on Pinedale and Bull Lake moraines, Willow Lake, Wind River Mountains, Wyoming. *Catena* 12, 329–42.

Tansley, A. G. (1935) The use and abuse of vegetational concepts and terms. *Ecology* 16, 284–307.

—— (1939) *The British Isles and Their Vegetation*. Cambridge: Cambridge University Press.

Tatsumi, Y. (2005) The subduction factory: how it operates in the evolving Earth. *GSA Today* 15(7), 4–10.

Taulman, J. F. and Robbins, L. W. (1996) Recent range expansion and distributional limits of the nine-banded armadillo (*Dasypus novemcinctus*) in the United Sates. *Journal of Biogeography* 23, 635–48.

Taylor, F. B. (1910) Bearing of the Tertiary mountain belt on the origin of the Earth's plan. *Bulletin of the Geological Society of America* 21, 179–226.

Temperton, V. M., Hobbs, R. J., Nuttle, T., and Halle, S. (2004) *Assembly Rules and Restoration Ecology*. Washington, DC: Island Press.

Terrill, C. (2007) *Unnatural Landscapes: Tracking Invasive Species* (Foreword by Gary Paul Nabhan). Tucson, AZ: University of Arizona Press.

Thomas, M. F. (1965) Some aspects of the geomorphology of tors and domes in Nigeria. *Zeitschrift für Geomorphologie* NF 9, 63–81.

Thompson, J. A. and Bell, J. C. (1998) Hydric conditions and hydromorphic properties within a Mollisol catena in southeastern Minnesota. *Soil Science Society of America Journal* 62, 1116–25.

Thompson, J. A., Bell, J. C., and Zanner, C. W. (1998) Hydrology and hydric soil extent within a Mollisol catena in southeastern Minnesota. *Soil Science Society of America Journal* 62, 1126–33.

Thompson, J. N. (1998) Rapid evolution as an ecological process. *Trends in Ecology and Evolution* 13, 329–32.

Thorn, C. E. and Welford, M. R. (1994) The equilibrium concept in geomorphology. *Annals of the Association of American Geographers* 84, 666–96.

Thornes, J. B. and Brunsden, D. (1977) *Geomorphology and Time.* London: Methuen.

Thuiller, W. (2003) BIOMOD – optimising predications of species distributions and projecting potential future shifts under global change. *Global Change Biology* 9, 1353–62.

Tobey, R. (1981) *Saving the Prairies: The Life Cycle of the Founding School of American Plant Ecology, 1895–1955.* Berkeley, CA: University of California Press.

Tooth, S. (2008) Arid geomorphology: recent progress from an Earth System Science perspective. *Progress in Physical Geography* 32, 81–101.

Travis, J. M. J. (2003) Climate change and habitat destruction: a deadly anthropogenic cocktail. *Proceedings of the Royal Society, London,* 270B, 467–73.

Troeh, F. R. (1964) Landform parameters correlated to soil drainage. *Soil Science Society of America Proceedings* 28, 808–12.

Twidale, C. R. (2002) The two-stage concept of landform and landscape development involving etching: origin, development and implications of an idea. *Earth-Science Reviews* 57, 37–74.

Vaughan, T. A. (1978) *Mammalogy,* 2nd edn. Philadelphia, PA: W. B. Saunders.

Verboom, J., Schotman, A., Opdam, P., and Metz, J. A. J. (1991) European nuthatch metapopulations in a fragmented agricultural landscape. *Oikos* 61, 149–56.

Verschuur, G. L. (1998) *Impact!: The Threat of Comets and Asteroids.* Oxford: Oxford University Press.

Via, S. (2001) Sympatric speciation in animals: the ugly duckling grows up. *Trends in Ecology and Evolution* 16, 381–90.

Volk, T. (2007) The properties of organisms are not tunable parameters selected because they create maximum entropy production on the biosphere scale. A byproduct framework in response to Kleidon. *Climatic Change* 85, 251–58.

Vreeken, W. J. (1973) Soil variability in small loess watersheds: clay and organic matter content. *Catena* 2, 321–36.

Waddington, C. H. (1957) *The Strategy of the Genes: A Discussion of Some Aspects of Theoretical Biology.* London: Macmillan.

Walker, L. R. and Moral, R. del (2003) *Primary Succession and Ecosystem Rehabilitation.* Cambridge: Cambridge University Press.

Walker, L. R., Walker, J., and Hobbs, R. J. (eds) (2007) *Linking Restoration and Ecological Succession.* New York: Springer.

Walter, H. and Lieth, H. (1960–67) *Klimadiagramm–Weltatlas.* Jena: Gustav Fischer.

Ward, P. D. (2007) *Under a Green Sky: Global Warming, the Mass Extinctions of the Past and What They Can Tell Us About Our Future.* New York: HarperCollins.

Watt, A. S. (1924) On the ecology of British beechwoods with special reference to their regeneration. II. The development and structure of beech communities on the Sussex Downs. *Journal of Ecology* 12, 145–204.

—— (1947) Pattern and process in the plant community. *Journal of Ecology* 35, 1–22.

Watts, A. B. (2001) *Isostasy and the Flexure of the Lithosphere.* Cambridge: Cambridge University Press.

WCED (1987) *Our Common Future.* Oxford: Oxford University Press for the UN World Commission on Economy and Environment.

Webb, N. R. and Thomas, J. A. (1994) Conserving insect habitats in heathland biotopes: a question of scale. In P. J. Edwards, R. M. May, and N. R. Webb (eds) *Large-Scale Ecology and Conservation Biology* (The 35th Symposium of the British Ecological Society with the Society for Conservation Biology, University of Southampton, 1993), pp. 129–51. Oxford: Blackwell Scientific Publications.

Wegener, A. L. (1912) Die Entstehung der Kontinente. *Petermanns Mitteilungen* 185–95, 253–56, 305–9.

—— (1915) *Die Entstehung der Kontinente und Ozeane.* Braunschweig: Friedrich Vieweg und Sohn.

—— (1966) *The Origin of Continents and Oceans.* Translated by J. Biram, with an introduction by B. C. King. London: Methuen.

Weiss, S. and Ferrand, N. (2007) *Phylogeography of Southern European Refugia: Evolutionary Perspectives on the Origins and Conservation of European Biodiversity.* Dordrecht, The Netherlands: Springer.

Whelan, R. J. (2008) *The Ecology of Fire.* Cambridge: Cambridge University Press.

Whewell, W. (1832) [Review of Lyell, 1830–33, vol. ii]. *Quarterly Review* 47, 103–32.

White, R. E. (2005) *Principles and Practice of Soil Science: The Soil as a Natural Resource,* 4th edn. Malden, MA: Blackwell.

Whitehead, A. N. (1925) *Science and the Modern World.* New York: The Free Press.

Whittaker, R. H. (1953) A consideration of climax theory: the climax as a population and pattern. *Ecological Monographs* 23, 41–78.

Whittaker, R. J. and Fernandez-Palacios, J. M. (2006) *Island Biogeography: Ecology, Evolution, and Conservation,* 2nd edn. Oxford: Oxford University Press.

Whittaker, R. J. and Jones, S. H. (1994) Structure in re-building insular ecosystems: an empirically derived model. *Oikos* 69, 524–30.

Whittaker, R. J., Bush, M. B., and Richards, K. (1989) Plant recolonization and vegetation succession on the Krakatau Islands, Indonesia. *Ecological Monographs* 59, 59–123.

Whittaker, R. J., Bush, M. B., Asquith, N. M., and Richards, K. (1992) Ecological aspects of plant colonisation of the Krakatau Islands. *GeoJournal* 28, 201–11.

WHO (World Health Organization) (1997) El Niño and its health impacts. *Journal of Communicable Diseases* 29, 375–77.

Wiener, N. (1948) *Cybernetics; or, Control and Communication in the Animal and the Machine.* Paris: Hermann et Cie; New York: The Technology Press.

Wiens, J. A., Moss, M. R., Turner, M. G., and Mladenoff, D. J. (eds) (2006) *Foundation Papers in Landscape Ecology.* New York: Columbia University Press.

Wigley, T. M. L. and Kelly, P. M. (1990) Holocene climatic change, ^{14}C wiggles and variations in solar irradiance. *Philosophical Transactions of the Royal Society of London* 330A, 547–60.

Willett, S. D., Hovius, N., Brandon, M. T., and Fisher, D. M. (eds) (2006) *Tectonics, Climate, and Landscape Evolution* (Geological Society of America Special Paper 398). Boulder, CO: The Geological Society of America.

Williams, G. C. (1992) *Natural Selection: Domains, Levels, and Challenges* (Oxford Series in Ecology and Evolution, Vol. 4). New York and Oxford: Oxford University Press.

Williams, J. W. and Jackson, S. T. (2007) Novel climates, no-analog communities, andecological surprises. *Frontiers in Ecology and the Environment* 5, 475–82.

Williams, J. W., Jackson, S. T., and Kutzbach, J. E. (2007) Projected distributions of novel and disappearing climates by 2100 AD. *Proceedings of the National Academy of Sciences USA* 104, 5, 738–42.

Williams, M. A. J. (1968) Termites and soil development near Brocks Creek, Northern Australia. *Australian Journal of Soil Science* 31, 153–54.

Willis, A. J. (1994) Arthur Roy Clapham, 1904–90. *Biographical Memoirs of Fellows of the Royal Society* 39, 73–90.

—— (1997) The ecosystem: an evolving concept viewed historically. *Functional Ecology* 11, 268–71.

Willis, K. J. and Whittaker, R. J. (2000) The refugial debate. *Science* 287, 1406–7.

Willis, K. J., Rudner, E., and Sümegi, P. (2000) The full-glacial forests of central and southeastern Europe. *Quaternary Rresearch* 53, 203–13.

Willmer, P., Stone, G., and Johnston, I. (2004) *Environmental Physiology of Animals*, 2nd edn. Oxford: Blackwell.

Wilson, E. O. (1988) *Biodiversity* (Papers From the First National Forum on BioDiversity (sic), September 1986, Washington, DC). Washington, DC: National Academy Press.

Wilson, E. O. and Forman, R. T. T. (2008) *Land Mosaics: The Ecology of Landscapes and Regions*. Cambridge: Cambridge University Press.

Wilson, J. T. (1963) A possible origin of the Hawaiian Islands. *Canadian Journal of Physics* 41, 8632–70.

Womack, W. R. and Schumm, S. A. (1977) Terraces of Douglas Creek, northwestern Colorado: an example of episodic erosion. *Geology* 5, 72–76.

Woodroffe, C. D. (2007) The natural resilience of coastal systems: primary concepts. In L. McFadden, R. J. Nicholls, and E. Penning-Rowsell (eds) *Managing Coastal Vulnerability*, pp. 45–60. Oxford and Amsterdam: Elsevier.

Woods, M. and Moriarty, P. V. (2001) Strangers in a strange land: the problem of exotic species. *Environmental Values* 10, 163–91.

Wunsch, C. (2004) Quantitative estimate of the Milankovitch-forced contribution to observed Quaternary climate change. *Quaternary Science Reviews* 23, 1001–12.

Xu, T., Moore, I. D., and Gallant, J. C. (1993) Fractals, fractal dimensions and landscapes – a review. *Geomorphology* 8, 245–62.

Yndestad, H. (2006) The influence of the lunar nodal cycle on Arctic climate. *ICES Journal of Marine Science* 63, 401–20.

Yuen, D. A., Maruyama, S., Karato, S.-I., and Windley, B. F. (2007) *Superplumes: Beyond Plate Tectonics*. Dordrecht, The Netherlands: Springer.

INDEX

Main entries are **emboldened**

active margins **3–6**
actualism **7–9**, 74, 175
adaptation **9–11**, 13, 55, 79, 122, 150, 152
adaptive radiation **11–13**
advection **13–14**, 45, 59
allogenic succession 156
allopatric speciation 11, 137, 150–2
aridity **14–15**, 89; index 14–15
astronomical (orbital) forcing **15–17**, 46, 96
Atlantic-type margins *see* passive margins
autogenic succession 155–6

bioaccumulation **17–18**
bioclimate **18–19**
bioclimatology 18
biodiversity **19–21**; and geodiversity 92; loss **19–21**, 107, 108; and refugia 137; and sustainability 156
biogeochemical cycles **21–4**; and climate change 35; and Gaia 83
biogeographical rules 10
biomagnification **17–18**
bombardment **24–6**; and catastrophism 27–8; and disturbance 53
Butterfly Effect 42

carbon cycle 22–3, 35
carrying capacity **26–7**; and desertification 47; and island biogeography 112; and sustainability 157
catastrophism 7, **27–8**, 100–1, 176; and bombardment 25–6
catena **28–30**; geomorphic 30; and slopes 172; and soil–landscapes 146–7
chaos 40–2, 61, 85
chronosequence **30–3**, 82; and ergodicity 66–8
climate change **33–6**; abrupt 167; and astronomical forcing 16; and biodiversity loss 21; and extinction 78–9; and geomorphic systems 166; and global warming 95–9; and species ranges 17–18; and solar forcing 148
climatic normals 60
climax community **37–8**, 72
cline 10, 152
'clorpt' equation 81–3
community change **38–40**; and exotic species 109 *see also* succession
complexification 72
complexity **40–4**, 66, 81, 160, 169
continental drift **44–5**; and mass